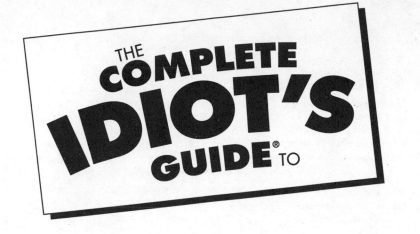

THE
COMPLETE IDIOT'S GUIDE® TO

Statistics

by Robert A. Donnelly Jr., Ph.D.

ALPHA

A member of Penguin Group (USA) Inc.

To my wife, Debbie, who supported and encouraged me every step of the way. I could not have done this without you, Babe.

International Standard Book Number: 1-59257-199-9
Library of Congress Catalog Card Number: 2004100485

06   05   04   8   7   6   5   4   3   2   1

Interpretation of the printing code: The rightmost number of the first series of numbers is the year of the book's printing; the rightmost number of the second series of numbers is the number of the book's printing. For example, a printing code of 04-1 shows that the first printing occurred in 2004.

Printed in the United States of America

Note: This publication contains the opinions and ideas of its author. It is intended to provide helpful and informative material on the subject matter covered. It is sold with the understanding that the author and publisher are not engaged in rendering professional services in the book. If the reader requires personal assistance or advice, a competent professional should be consulted.

The author and publisher specifically disclaim any responsibility for any liability, loss, or risk, personal or otherwise, which is incurred as a consequence, directly or indirectly, of the use and application of any of the contents of this book.

Most Alpha books are available at special quantity discounts for bulk purchases for sales promotions, premiums, fund-raising, or educational use. Special books, or book excerpts, can also be created to fit specific needs.

For details, write: Special Markets, Alpha Books, 375 Hudson Street, New York, NY 10014.

Publisher: *Marie Butler-Knight*
Product Manager: *Phil Kitchel*
Senior Managing Editor: *Jennifer Chisholm*
Senior Acquisitions Editor: *Mike Sanders*
Development Editor: *Nancy D. Lewis*
Senior Production Editor: *Christy Wagner*
Production Editor: *Megan Douglass*
Copy Editor: *Keith Cline*
Illustrator: *Chris Eliopoulos*
Cover/Book Designer: *Trina Wurst*
Indexer: *Tonya Heard*
Layout/Proofreading: *John Etchison, Becky Harmon*

Contents at a Glance

Appendixes

Contents

Appendixes

Foreword

Statistics, statistics everywhere, but not a single word can we understand! Actually, understanding statistics is a critically important skill that we all need to have in this day and age. Every day, we are inundated with data about politics, sports, business, the stock market, health issues, financial matters, and many other topics. Most of us don't pay much attention to a lot of the statistics we hear, but more importantly, most of us don't really understand how to make sense of the numbers, ratios, and percentages with which we are constantly barraged. In order to obtain the truth behind the numbers, we must be able to ascertain what the data is really saying to us. We need to determine whether the data is biased in a particular direction, or whether the true, balanced picture is correctly represented in the numbers. That is the reason for reading this book.

Statistics, as a field, is usually not the most popular topic or course in school. In fact, many people will go to great lengths to avoid having to take a statistics course. Many people think of it as a math course, or something that is very quantitative, and that scares them away. Others, who get past the math, do not have the patience to search for what the numbers are actually saying. And still others don't believe that statistics can ever be used in a legitimate manner to point to the truth. But whether it is about significant trends in the population, average salary and unemployment rates, or similarities and differences across stock prices, statistics are an extremely important input to many decisions that we are faced with. And understanding how to generate the statistics and interpret them relating to your particular decision can make all the difference between a good decision and a poor one.

For example, suppose that you are trying to sell your house and you need to set a selling price for it. The mean selling price of houses in your area is $250,000, so you set your price at $265,000. Perhaps $250,000 is the price roughly in the middle of several house prices that have ranged from $200,000 to $270,000, so you are in the ballpark. However, a mean of $250,000 could also occur with house prices of $175,000, $150,000, $145,000, $100,000, and $780,000. One high price out of five causes the mean to increase dramatically, so you have potentially priced yourself out of most of the market. For this reason, it is important to understand what the term "mean" really represents.

Another compelling reason to understand statistics is that we are living in a quality-driven society. Everything nowadays is related to "improving quality," a "quality job," or "quality improvement processes." Companies are striving for higher quality in their products and employees, and are using such programs as "continuous improvement" and "six-sigma" to achieve and measure this quality. Even the ordinary consumer has heard these terms and needs to understand them in order to be an educated customer or client. Here again, an understanding of statistics can help you make wise choices related to purchasing behavior.

So as we move from the information age to the knowledge age, it is becoming increasingly important for us to at least understand, if not generate and use, statistics. Bob Donnelly has done a wonderful job of presenting statistics in this book so that you can improve your ability to look at and comprehend the data you run across every day. Bob's many years of teaching statistics at all levels have provided the basis for his phenomenal ability to explain difficult statistical concepts clearly. Even the most unsophisticated reader will soon understand the subtleties and power of telling the truth with statistics!

Christine T. Kydd
2003 Delaware Professor of the Year
Associate Professor of Business Administration and
Director of Undergraduate Programs
University of Delaware

Introduction

Statistics. Why does this single word terrify so many of today's students? The mere mention of this word in the classroom causes a glassy-eyed, deer-in-the-headlights reaction across a sea of faces. In one form or another, the topic of statistics has been torturing innocent students for hundreds of years. You would think the word statistics had been derived from the Latin words *sta*, meaning "Why" and *tistica*, meaning "do I have to take this %#!@*% class?" It really doesn't have to be this way. The term "stat" needn't be a four-letter word in the minds of our students.

As you read this paragraph, you're probably wondering what can this book do for you. Well, it's written by a person (that's me) who (a) clearly remembers being in your shoes as a student (even if it was in the last century), (b) sympathizes with your current dilemma (I can feel your pain), and (c) has learned a thing or two over many years of teaching (those many hours of tutorials were not for naught). The result of this experience has allowed me to discover ways to walk you through many of the concepts that traditionally frustrate students. Armed with the tools that you will gain from the many examples and numerous problems explained in detail, this task will not be as daunting as it first appears.

Unfortunately, fancy terms such as inferential statistics, analysis of variance, and hypothesis testing are enough to send many running for the hills. My goal has been to show that these complicated terms are really used to describe ordinary, straight-forward concepts. By applying many of the techniques to everyday (and sometimes humorous) examples, I have attempted to show that not only is statistics a topic that can be mastered by anyone, it can actually make sense and be helpful in numerous situations.

To further help those in need, I have established a companion website for this book at www.stat-guide.com. Here you will find additional problems with solutions and links to other useful websites. If you have any feedback you would like to provide about this book, please send me an e-mail via this website.

So hold on to your hats, we're about to take a wild ride into the realm of numbers, inequalities, and, oh yes, don't forget all those Greek symbols! You will see equations that look like the Chinese alphabet at first glance, but can, in fact, be simplified into plain English. The step-by-step description of each problem will help you break down the process into manageable pieces. As you work the example problems on your own, you will gain confidence and success in your abilities to put numbers to work to provide usable information. And, guess what, that is sometimes how statisticians are born!

How This Book Is Organized

The book is organized into four parts:

In **Part 1, "The Basics,"** we start from the very beginning without any assumptions of prior knowledge. After a brief history lesson to warm you up, we dive into the world of data and learn about the different types of data and the variety of measurement scales that we can use. We also cover how to display data graphically, both manually and with the help of Microsoft Excel. We wrap up Part 1 with learning how to calculate descriptive statistics of a sample, such as the mean and standard deviation.

In **Part 2, "Probability Topics,"** we introduce the scary world of probability theory. Once again, I assume you have no prior knowledge of this topic (or if you did, I assume you buried it in the deep recesses of your brain, hoping to *never* uncover it). An important topic in this section is learning how to count the number of events, which can really improve your poker skills. After easing you into the basics, we gently slide into probability distributions, such as the normal and binomial. Once these are mastered, we have set the stage for Part 3.

In **Part 3, "Inferential Statistics,"** we start off learning about sampling procedures and the way samples behave statistically. When these concepts are understood, we start acting like real statisticians by making estimates of populations using confidence intervals. By this time, your own mother wouldn't recognize you! We'll top Part 3 off with a procedure that's near and dear to every statistician's heart—hypothesis testing. With this tool, you can do things like make bold comparisons between the male and female population. I'll leave that one to you.

In **Part 4, "Advanced Inferential Statistics,"** we build on earlier topics and explore analysis of variance, a popular method to compare more than two populations to each other. We will also learn about the chi-square tests, which enable us to determine whether two variables are dependent. And last but not least, we'll discover how simple regression (which, by the way, is not so simple or else it wouldn't be the last topic in the book) describes the strength and direction of the relationship between two variables. When you're done with these topics, your friends won't believe the words they hear coming from your mouth.

Extras

Throughout this book, you will come across various sidebars that provide a helping hand when things seem to get a little tough. Many are based on my experience as a teacher with the concepts that I have found to cause students the most difficulty.

Stat Facts

These are definitions of statistical jargon explained in a nonthreatening manner, which will help to clarify important concepts. You'll find that their bark is often far worse than their bite.

Random Thoughts

Insights that I find interesting (and hopefully you will, too!) about the current topic. Statistics is full of little-known facts that can help relieve the intensity of the topic at hand.

Bob's Basics

These are tips and insights that I have accumulated over the years of helping students master a particular topic. The goal here is to have that light bulb in the brain go off, resulting in the feeling of "I got it!"

Wrong Number

These are warnings of potential pitfalls that are lying in wait for an unsuspecting student to fall into. By taking note of these, you'll avoid the same traps that have ensnarled many of your predecessors.

Acknowledgments

There are many people whom I am indebted to for helping me with this project. I'd like to thank Jessica Faust for her guidance and expertise to get me on track in the beginning, Mike Sanders for going easy on me with his initial feedback, and Nancy Lewis, for her valuable opinions during the writing process.

To my colleague and friend, Dr. Patricia Buhler, who introduced me to the publishing industry, convinced me to take on this project, and encouraged me throughout the writing process. This all started with you, Pat.

To my in-laws, Lindsay and Marge, who never failed to ask me what chapter I was writing, which motivated me to stay on schedule. Your commitment to each other is a true inspiration for all of us.

To my boss of 10 years at Goldey-Beacom College, Joyce Jones, who rearranged my teaching schedule to accommodate my deadlines. Life at GBC will never be the same after you retire, Joyce. I am really going to miss you. Thank you for your constant support over the years. You have been a great boss and a true friend.

To my students who make teaching a pleasure. The lessons that I have leaned over the years about teaching were invaluable to me as I wrote this book. Without all of you, I would never have had the opportunity to be an author.

xx The Complete Idiot's Guide to Statistics

To my children, Christin, Brian, and John, and my stepchildren, Katie, Sam, and Jeff, for your interest in this book and your willingness to let me use your antics as examples in many of the chapters.

And most importantly, to my wife, Debbie, who made this a team effort with all the hours she spent contributing ideas, proofreading manuscripts, editing figures, and giving up family time to help me stay on schedule. Deb's excitement over the opportunity to write this book gave me the courage to accept this challenge. Deb was also the inspiration for many of the examples used in the book, allowing me to share experiences from our wonderful life together. Thank you for your love and your patience with me while writing this book.

Trademarks

All terms mentioned in this book that are known to be or are suspected of being trademarks or service marks have been appropriately capitalized. Alpha Books and Penguin Group (USA) Inc. cannot attest to the accuracy of this information. Use of a term in this book should not be regarded as affecting the validity of any trademark or service mark.

Part 1

The Basics

The key to successfully mastering statistics is to have a solid foundation of the basics. To get a firm grasp of the more advanced topics, you need to be well grounded in the concepts presented in this part. After a quick history lesson, the following chapters focus on data, the starting point for any method in statistics. You might be surprised with how much there really is to learn about data and all of its properties. We will examine the different types of data, how it is collected, how it is displayed, and how it is used to calculate things called the mean and standard deviation.

Chapter 1

Let's Get Started

In This Chapter

- The purpose of statistics—what's in it for you?
- The history of statistics—where did this stuff come from?
- Brief overview of the field of statistics
- The ethical side of statistics

How often have you asked yourself why you even need to learn statistics? Well, you're not alone. All too often students find themselves drowning in a mathematical swamp of theories and concepts and never get a chance to see the "big picture" before going under. My goal in this chapter is to provide you with that broader perspective and convince you that statistics is a very useful tool in our current society. In other words, here comes your life preserver. Grab on!

In today's technologically advanced world, we are surrounded by a barrage of data and information from sources that are trying to convince us to buy something or simply persuade us to agree with their point of view. When we hear on TV that a politician is leading in the polls and in small print we see + or – 4 percent, do we know what that means? When a new product is recommended by 4 out of 5 doctors, do we question the validity of the claim? (For instance, were the doctors paid for their endorsement?)

Statistics can have a powerful influence on our feelings, our opinions, and our decisions that we make in life. Getting a handle on this widely used tool is a good thing for all of us.

Where Is This Stuff Used?

The *Funk and Wagnalls Dictionary* I just reached for on my bookshelf defines *statistics* as "the science that deals with the collection, tabulation, and systematic classification of quantitative data, especially as a basis for inference and induction." Now that's a mouthful. In simpler terms, I view statistics as a way to take numbers and convert them into useful information so that good decisions can be made.

These decisions can affect our lives in many ways. For instance, countless medical studies have been performed to determine the effectiveness of new drugs. Statistics form the basis of making an objective decision as to whether this new drug is actually an improvement over current treatments. Government policies are often dictated by the results of statistical studies and by the manner in which these results are presented.

CAUTION

Wrong Number

Not interpreting statistical information properly can lead to disaster. Coca-Cola performed a major consumer study in 1985 and, based on the results, decided to reformulate Coke, its flagship drink. After a huge public outcry, Coca-Cola had to backtrack and bring the original formulation back to market. What a mess!

Today's corporations are making major business decisions based on statistical analysis. In the 1980s, Marriott conducted an extensive survey with potential customers on their attitudes about current hotel offerings. After analyzing the data, Courtyard by Marriott was launched and has been a huge success for the company.

The federal government heavily relies on the national census that is conducted every 10 years to determine funding levels for all the various parts of the country. The statistical analysis that is performed on this census data has far-reaching implications for many ongoing programs at the state and federal levels.

The entire sports industry is completely dependent on the field of statistics. Can you even imagine baseball, football, or basketball without all the statistical analysis that surrounds them? You would never know who the top players are, who is currently hot, and who is in a slump. But then, without statistics, how could the players negotiate these outrageous salaries? Hmmm, maybe I'm onto something here.

My point in this section is to raise your awareness to the fact that we are surrounded by statistics in our society and that our world would be very different if this wasn't the case. Statistics is a useful, and sometimes even critical, tool in our everyday life.

Who Thought of This Stuff?

The field of statistics has been evolving for a very long time. Population surveys appear to be the primary motivation for the historical development of statistics as we know it today. In fact, according to the Bible, the Roman Empire conducted a census more than 2,000 years ago. The very word "statistics" itself comes from the Latin word *status*, which means "state." This etymological connection reflects the earliest focus of statistics on measuring things such as the number of (taxable) subjects in a kingdom (or state) or the number of subjects to send off to invade neighboring kingdoms.

Early Pioneers

European mathematicians provided the basic foundation for the field of statistics. In 1532, Sir William Petty provided the first accounts of the number of deaths in London on a weekly basis. So begins the insurance companies' morbid fascination with death statistics.

During the 1600s, Swiss mathematician James Bernoulli is credited with calculating the probability of a sequence of events, otherwise known as "independent trials." This is an unfortunate choice of words; many students over the generations have struggled with this concept and have felt like they were on "trial" themselves. You might remember dealing with the problem of calculating (or trying desperately to calculate) the probability of 7 "heads" in 10 coin tosses in a math class. You can thank Mr. Bernoulli for providing you with a way to solve this type of problem. Chapter 9 explores Bernoulli trials in loving detail, and with a little practice you'll get off with a light sentence.

English mathematician Thomas Bayes developed probability concepts during the 1700s that have also been very useful to the field of statistics. Bayes used probability of known events of the past to predict probabilities of the future. This concept of *inference* is widely used in statistical techniques today. Chapter 7 covers one of his particular contributions, appropriately known as Bayes' theorem.

Stat Facts _____

The term **inference** refers to a key concept in statistics in which we make a conclusion from available evidence.

More Recent Famous People

It wasn't until the early twentieth century that statistics began to develop into the field that we know it as today. William Gossett developed the famous "t-test" using the

Student-t probability distribution while working at the Guinness brewery in Dublin, Ireland. We will raise our glasses to Mr. Gossett as we investigate his efforts in Chapter 14.

W. Edwards Deming has been credited with merging the science of statistics with the field of quality control in manufacturing environments. Dr. Deming spent considerable time in Japan during the 1950s and 1960s promoting the concept of statistical quality for businesses. This technique relies on control charts to monitor a process and the use of statistics to determine whether the process is operating satisfactorily. During the 1970s, the Japanese auto industry gained major market share in this country mainly due to superior quality. That's the power of statistics!

> ### Random Thoughts
>
> Dr. Deming's philosophy has been condensed down to what is known as Deming's 14 points. This list has proved to be invaluable for organizations that are seeking ways to use statistics to make their processes more efficient. Through Dr. Deming's efforts, statistics has found a significant role in the business world. Check out his book *The Deming Management Method* (Perigee, 1988) for more information.

The Field of Statistics Today

The science of statistics has evolved into two basic categories known as *descriptive* statistics and *inferential* statistics. Because descriptive statistics is generally simpler, it can be thought of as the "minor league" of the field; whereas inferential statistics, being more challenging, can be considered the "major league" of the two.

Stat Facts

The purpose of **descriptive** statistics is to summarize or display data so that we can quickly obtain an overview. **Inferential** statistics allows us to make claims or conclusions about a population based on a sample of data from that population. A population represents all possible outcomes or measurements of interest. A sample is a subset of a population.

Today, computers and software play a dominant role in our use of statistics. Current desktop computers have the capability to process and analyze huge amounts of data and information. Specialized software such as SAS and SPSS allow you to conveniently perform all sorts of complicated statistical techniques without breaking a sweat.

In this book, I will show you how to perform many statistical techniques using Microsoft Excel, a spreadsheet software package that's readily available on most desktop computers (also included in the Microsoft

Office software suite). Excel has many easy-to-use statistics features that can save you time and energy. If this paragraph causes your blood pressure to elevate (hey, wait a minute, nobody told me this was a computer book!), have no fear. Feel free to just skip over these sections; subsequent material in this book does not depend on this information. I promise it will not be on the final exam.

Descriptive Statistics—the Minor League

The main focus of descriptive statistics is to summarize and display data. Descriptive statistics plays an important role today because of the vast amount of data readily available at our fingertips. With a basic computer and an Internet connection, we can access volumes of data in no time at all. Being able to accurately summarize all of this data to get a look at the "big picture," either graphically or numerically, is the job of descriptive statistics.

There are many examples of descriptive statistics, but the most common is the average. As an example, let's say I would like to get a perspective on the average attention span of my Labrador retriever by using flash cards. I time each incident with a stopwatch and write it down on my clipboard. The following table lists our results, measured in seconds:

Observation	Seconds
1	4
2	8
3	5
4	10
5	2
6	4
7	7
8	12
9	7

Using descriptive statistics, I can calculate the average attention span, as follows:

$$\frac{4+8+5+10+2+4+7+12+7}{9} = 6.6 \text{ seconds}$$

Descriptive statistics can also involve displaying the data graphically, as shown Figure 1.1. What a good dog!

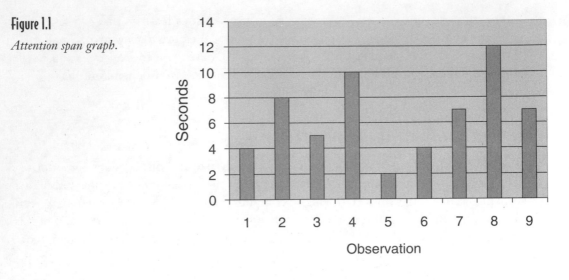

Figure 1.1

Attention span graph.

We will delve into descriptive statistics in more detail in Chapters 3 and 4. But until then, we're ready to move up to the big leagues—inferential statistics.

Inferential Statistics—the Major League

As important as descriptive statistics is to us number crunchers, what we really get excited about is inferential statistics. This category covers a large variety of techniques that allow us to make actual claims about a population based on a sample of data. Suppose, for instance, that I am interested in discovering in general who has the longer attention span, Labrador retrievers or, let's say, teenage boys. (Based on personal observations, I suspect I know the answer to this, but I'll keep it to myself.) Now, it's not possible to measure the attention span of every teenager and every dog, so the next best thing is to take a sample of each and measure them.

At this point, I need to explain the difference between a population and a sample. The term "population" is used in statistics to represent all possible measurements or outcomes that are of interest to us in a particular study. The term "sample" refers to a portion of the population that is representative of the population from which it was selected.

In this example, the population is all teenage boys and all Labrador retrievers. I need to select a sample of teenagers and a sample of dogs that represent their respective populations. Based on the results of my samples, I can *infer* the average attention span of each population and determine which is longer.

Figure 1.2 shows the relationship between a population and a sample.

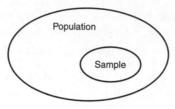

Figure 1.2

The relationship between a population and a sample.

The following are other examples of inferential statistics:

◆ Based on a recent sample, I am 95 percent certain that the average age of my customers is between 32 and 35 years old.

◆ The average salary for male employees in a particular job category across the country was higher than the female employees' salary, based on a random survey.

◆ Median weekly earnings across the country fell 1.5 percent in the first quarter of 2003. Ouch!!! (*Time* magazine, May 26, 2003, page 46)

In each case, the findings were based on a sample from a larger population and used to make an inference on that entire population.

The basic difference between descriptive and inferential statistics is that descriptive statistics reports on only the observations at hand and nothing more. Inferential statistics makes a statement about a population based solely on results from a sample taken from that population.

I feel compelled to tell you at this point that inferential statistics is the area of this field that students find the most challenging. To be able to make statements based on samples, mathematical models need to be used that involve probability theory. Now don't panic. Take a deep breath and count to 10 slowly. That's better. I realize that this is often the stumbling block for many, so I have devoted plenty of pages to that nasty "p" word.

Bob's Basics

A good understanding of probability concepts is an essential stepping-stone for properly digesting statistics. Part 2 of this book adequately covers probability.

Ethics and Statistics—It's a Dangerous World out There

People often use statistics when attempting to persuade you to their point of view. They might be motivated to convince you to purchase something from them or simply to support them. This motivation can lead to the misuse of statistics in several ways.

One of the most common misuses is choosing a sample that ensures results consistent with the desired outcome, rather than choosing a sample representative of the population of interest. This is known as having a *biased sample*.

Suppose, for instance, that I'm an upstanding politician whose only concern is the best interest of my constituents and I want to propose that Congress establish a national golf holiday. During this honored day, all government and business offices would be closed so that we all could run out to chase a little white ball into a hole that's way too small with sticks purposely designed by the evil golf companies to make this task impossible. Sounds like fun to me! Somehow, I would need to demonstrate that the average level-headed American is in favor of this. Here is where the genius part of my plan lies: Rather than survey the general American public, I pass out my survey form only at golf courses. But wait … it only gets better. I design the survey to look like the following:

Stat Facts

A **biased sample** is a sample that does not represent the intended population and can lead to distorted findings. Biased sampling can occur either intentionally or unintentionally.

We would like to propose a national golf holiday, on which everybody gets the day off from work and plays golf *all day*. (This means you would not need permission from your spouse.) Are you in favor of this proposal?

A. Yes, most definitely
B. Sure, why not
C. No, I would rather spend the entire day at work

P.S. If you choose C, we will permanently revoke all of your golfing privileges everywhere in the country for the rest of your life. We are dead serious.

I can now honestly report back to Congress that the respondents of my survey where overwhelmingly in favor of this new holiday. And for what we know about Congress, they'd probably believe me.

Another way to misuse statistics is to make differences seem greater than they actually are by graphically presenting the data in a deceptive manner. Now that I have golf on my brain, let me use my golf scores to demonstrate this point. Let's say, hypothetically speaking, of course, that my average golf score during the month of May was 98. After taking some lessons in June, my average score in July dropped to 96. (For you non-golfers, lower is better.) The graph in Figure 1.3 shows that this improvement was nothing to write home about.

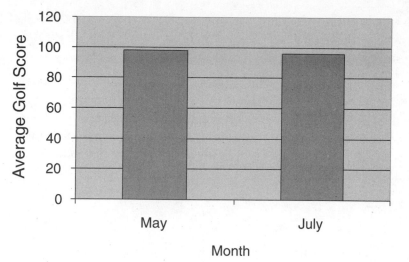

Figure 1.3

This graph shows the actual difference between May and July.

However, to avoid feeling like I wasted my money on lessons, I can present the difference between May and July on a different scale, as in Figure 1.4.

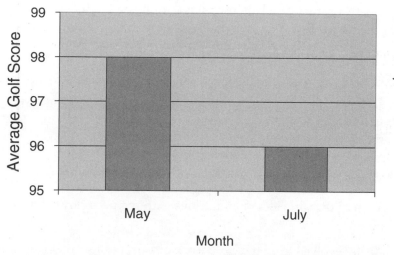

Figure 1.4

This graph exaggerates the difference between May and July.

By changing the scale of the graph, it appears that I really made progress on my golf game—when in reality, little progress was made. Oh well, back to the drawing board.

Many of the polls we see on the Internet represent another potential misuse of statistics. Many websites encourage visitors to vote on a question of the day. The results of these informal polls are unreliable simply because those collecting the data have no control over who responds or how many times they respond. As stated earlier, a valid statistical study depends on selecting a sample representative of the population of interest.

This is not possible when any person surfing the Internet can participate in the poll. Even though most of these polls state that the results are not scientific, it's a natural human tendency to still be influenced by the results that we see.

The lesson here is that we are all consumers of statistics. We are constantly surrounded by information provided by someone who is trying to influence us or gain our support. By having a basic understanding about the field of statistics, we increase the likelihood that we can ward off those evil spirits in their attempts to distort the truth. Starting with the next chapter, we'll begin our journey to achieve this goal … oh, and to help you pass your statistics course.

Your Turn

Identify each of the following statistics as either descriptive or inferential.

1. In 2001, 72.7 percent of Asian American households in the United States owned a computer. (*The News Journal*, May 21, 2003, page A1)

2. Households with children under the age of 18 are more likely to have access to the Internet (62 percent) than family households with no children (53 percent). ("A Nation Online," February 2002, www.ntia.doc.gov/ntiahome/dn/index.html)

3. Barry Bonds's lifetime batting average is .295. (www.espn.com)

4. The average SAT score for incoming freshman at a local college in 2002 was 550.

5. On a recent poll, 67 percent of Americans had a favorable opinion of President George W. Bush. (*Time* magazine, April 7, 2003, page 40)

The Least You Need to Know

◆ Statistics is a vital tool that provides organizations with the necessary information to make good decisions.

◆ The field of statistics evolved from the early work of European mathematicians during the seventeenth century.

◆ Descriptive statistics focuses on the summary or display of data so that we can quickly obtain an overview.

◆ Inferential statistics allows us to make claims or conclusions about a population based on a sample of data from that population.

◆ We are all consumers of statistics and need to be aware of the potential misuses that can occur in this field.

Data, Data Everywhere and Not a Drop to Drink

In This Chapter

- The difference between data and information
- Where does data come from?
- What kinds of data can be used?
- Different ways of measuring data
- Setting up Excel for statistical analysis

Data is the basic foundation for the field of statistics. The validity of any statistical study hinges on the validity of the data from the beginning of the process. Many things can come into question, such as the accuracy of the data or the source of the data to name a few. Without the proper foundation, your efforts to provide a sound analysis will come tumbling down.

The issues surrounding data can be surprisingly complex. After all, aren't we just talking about numbers here? What could go wrong? Well, plenty can. Data can be classified in several ways. We need to recognize the difference between quantitative and qualitative data and how each is used. Data can also be measured in many ways. The data measurement choice that we make at the start of the study will determine what kind of statistical techniques we can apply.

The Importance of Data

Data is simply defined as the value assigned to a specific observation or measurement. If I'm collecting data on my wife's snoring behavior, I can do so in different ways. I can measure how many times Debbie snores over a 10-minute period. I can measure the length of each snore in seconds. I could also measure how loud each snore is with a descriptive phrase, like "That one sounded like a bear just waking up from hibernation" or "Wow! That one sounded like an Alaskan seal calling for its young." (How a sound like that can come from a person who can fit into a pair of size 2 jeans and still be able to breathe I'll never know.)

In each case, I'm recording data on the same event in a different form. In the first case, I'm measuring a frequency, or number of occurrences. In the second instance, I'm measuring duration, or length in time. And the final attempt measures the event by describing volume using words rather than numbers. Each of these cases just shows a different way to use data.

If you haven't noticed yet, statistics people like to use all sorts of jargon. Here are a couple more terms. Data that is used to describe something of interest about a population is called a *parameter*. However, if the data is describing a sample from that population, we refer to it as a *statistic*. For instance, let's say that the population of interest is my wife's 3-year-old preschool class and my measurement of interest is how many times the little urchins use the bathroom in a day (according to Debbie, much more than should be physically possible).

We could average the number of trips per child, and this figure would be considered a parameter because the entire population was measured. However, if we want to make a statement about the average number of bathroom trips per day per 3 year old in the country, then Debbie's class could be our sample. The average that we observe from her class could be considered a statistic if we assume it could be used to estimate all 3 year olds in the country.

Stat Facts _____

Data is the value assigned to an observation or a measurement and is the building blocks to statistical analysis. The plural form is data and the singular form is datum, referring to an individual observation or measurement.

Data that describes a characteristic about a population is known as a **parameter**. Data that describes a characteristic about a sample is known as a **statistic**. **Information** is data that is transformed into useful facts that can be used for a specific purpose, such as making a decision.

Data is the building blocks of all statistical studies. You can hire the most expensive, well-known statisticians and provide them with the latest computer hardware and software available, but if the data you provide them is inaccurate or not relevant to the study, the final results will be worthless.

However, data all by its lonesome is not all that useful. By definition, data is just the raw facts and figures that pertain to a measurement of interest. *Information*, on the other hand, is derived from the facts for the purpose of making decisions. One of the major reasons to use statistics is to transform data into information. For example, the table that follows shows monthly sales data for a small retail store.

Monthly Sales Data

Month	Sales ($)
January	15,178
February	14,293
March	13,492
April	12,287
May	11,321

Using statistical analysis, we can generate information that may be of interest, such as "Wake up! You are doing something very wrong. At this rate, you will be out of business by early next year." Based on this valuable information, we can now make some important decisions about how to avoid this impending disaster.

The Sources of Data—Where Does All This Stuff Come From?

The sources of data can be classified in two broad categories: *primary* and *secondary*. Secondary data is data that somebody else has collected and made available for others to use. The U.S. government loves to collect and publish all sorts of interesting data, just in case anyone should need it. The Department of Commerce handles census data, and the Department of Labor collects mountains of, you guessed it, labor statistics. The Department of the Interior provides all sorts of data about U.S. resources. For instance, did you know that there are 250 species of squirrels in this country? If you don't believe me, go to www.npwrc.usgs.gov/resource/distr/mammals/mammals/squirrel.htm and you can become the local "squirrel" expert.

Stat Facts

Primary data is data that you have collected for your own use. **Secondary data** is data collected by someone else that you are "borrowing."

The Canadian government has a great system for providing statistical data to the public. Rather than each department in the government being responsible for collecting and disbursing data as in the United States, Canada has a national statistical agency known as Statistics Canada (www.statcan.ca/start.html). It's like one-stop shopping for the statistician. It's a wonderful website that makes research of Canadian facts a pleasure.

The main drawback of using secondary data is that you have no control over how the data was collected. It's a natural human tendency to believe anything that's in print (You believe me, don't you?), and sometimes that requires a leap of faith. The advantage to secondary data is that it's cheap (sometimes free) and it's available now. That's called instant gratification.

Primary data, on the other hand, is data that is collected by the person who eventually uses this data. This type of data can be expensive to acquire, but the main advantage is that it's your data and you have nobody else to blame but yourself if you make a mess of it.

When collecting primary data, you want to ensure that the results will not be biased by the manner in which it is collected. There are many ways to obtain primary data, such as direct observation, surveys, and experiments.

Random Thoughts

The Internet has also become a rich source of data for statistics published by various industries. If you can muddle your way through the 63,278 sites that come back from the typical Internet search engine, you might find something useful. I once found a Japanese study on the effect of fluoride on toad embryos (www.fluoride-journal.com/ 1971.htm). Before this discovery, I was completely oblivious to the fact that toads even had teeth, much less a cavity problem. I can't wait to impress my friends at the next neighborhood dinner party.

Direct Observation—I'll Be Watching You

Most often, this method focuses on gathering data while the subjects of interest are in their natural environment, oblivious to what is going on around them. Examples of these studies would be observing wild animals stalking their prey in the forest or teenagers at the mall on Friday night (or is that the same example?). The advantage of this method is that the subjects will unlikely be influenced by the data collection.

Focus groups are a direct observational technique where the subjects are aware that data is being collected. Businesses use focus groups to gather information in a group setting that is controlled by a moderator. The subjects are usually paid for their time and are asked to comment on specific topics.

Experiments—Who's in Control?

This method is more direct than observation because the subjects will participate in an experiment designed to determine the effectiveness of a treatment. An example of a treatment could be the use of a new medical drug. Two groups would be established. The first is the experimental group who receive the new drug, and the second is the control group who think they are getting the new drug but are in fact getting no medication. The reactions from both groups are measured and compared to determine whether the new drug was effective.

The claims that the experimental studies are attempting to verify need to be clear and specific. I just recently read about an herb called ginkgo biloba. According to this article, people who make money selling funny-sounding herbs claim ginkgo biloba will keep your mind sharp as you age. Sounds like something everyone would want. Now let's see, where was I? As stated, this claim might prove difficult to verify. How do you define "keeping your mind sharp"? And then, how do you measure sharpness of mind? These are some of the challenges that statistical experiments face.

The benefit of experiments is that it allows the statistician to control factors that could influence the results, such as gender, age, and education of the participants. The concern about collecting data through experiments is that the response of the subjects could be influenced by the fact they are participating in a study. The design of experiments for a statistical study is a very complex topic and goes beyond the scope of this book.

Surveys—Is That Your Final Answer?

This technique of data collection involves directly asking the subject a series of questions. The questionnaire needs to be carefully designed to avoid any bias (see Chapter 1) or confusion for those participating. Concerns also exist about the influence the survey will have on the participant's responses. Some participants respond in a way they feel the survey would like them to. This is very similar to the manner in which hostages bond with their captors. The survey can be administered by e-mail, snail-mail, or telephone. It's the telephone survey that I'm most fond of, especially when I get the call just as I'm sitting down to dinner, getting in the shower, or finally making some progress on the chapter I'm writing.

Bob's Basics

Research has shown that the responses a person provides a questionnaire can be affected by the manner in which the questions are asked. A question that is posed in a positive tone will tend to invoke a more positive response and vice versa. A good strategy is to test your questionnaire with a small group of people before releasing it to the general public.

Whatever method you employ, your primary concern should always be that the sample is representative of the population in which you are interested.

Types of Data

Another way to classify data is by one of two types: *quantitative* or *qualitative*.

◆ Quantitative data uses numerical values to describe something of interest. An example is Debbie's age, which I have been forced by a legally bound document to never, never, never reveal anywhere in this book, not even if it's buried in an appendix as an answer to an obscure question. (*Hint:* See page 49.)

◆ Qualitative data uses descriptive terms to measure or classify something of interest. One example of qualitative data is the name of a respondent in a survey and his or her level of education. The next section covers qualitative data in more detail.

Types of Measurement Scales—a Weighty Topic

Who would have thought of so many ways to look at data? The final way to classify data is by the way it is measured. This distinction is critical because it affects which statistical techniques we can use in our analysis of the data. Each type is discussed in detail in the following sections.

Nominal Level of Measurement

A *nominal* level of measurement deals strictly with qualitative data. Observations are simply assigned to predetermined categories. One example is gender of the respondent, with the categories being male and female. Another example is data indicating the type of dog, if any, owned by families in a neighborhood. The categories for this data are the various dog types (black Lab, terrier, stupid mangy mutt that keeps me awake by barking all night at the moon). This data type does not allow us to perform any mathematical operations, such as adding or multiplying. We also cannot rank-order this list in any way from highest to lowest (although I would put black Lab at the top). This type is considered the lowest level of data and, as a result, is the most restrictive when choosing a statistical technique to use for the analysis.

Numbers can be used at the nominal level of measurement. Even in this case, the rules of the nominal scale still remain. An example would be zip codes or telephone numbers, which can't be added or placed in a meaningful order of greater than or less than. Even though the data appears to be numbers, it's handled just like qualitative data.

Ordinal Level of Measurement

On the food chain of data, *ordinal* is the next level up. It has all the properties of nominal data with the added feature that we can rank-order the values from highest to lowest. An example is if you were to have a lawnmower race. Let's say the finishing order was Scott, Tom, and Bob. We still can't perform mathematical operations on this data, but we can say that Scott's lawnmower was faster than Bob's. However, we cannot say how much faster. Ordinal data does not allow us to make measurements between the categories and to say, for instance, that Scott's lawnmower is twice a good as Bob's (it's not).

Ordinal data can be either qualitative or quantitative. An example of quantitative data is rating movies with 1, 2, 3, or 4 stars. However, we still may not claim that a 4-star movie is 4 times as good as a 1-star movie.

Interval Level of Measurement

Moving up the scale of data, we find ourselves at the *interval* level, which is strictly quantitative data. Now we can get to work with the mathematical operations of addition and subtraction when comparing values. For this data, the difference between the different categories can be measured with actual numbers and also provides meaningful information. Temperature measurement in degrees Fahrenheit is a common example here. For instance, 70 degrees is 5 degrees warmer than 65 degrees. However, multiplication and division can't be performed on this data. Why not? Simply because we cannot argue that 100 degrees is twice as warm as 50 degrees.

Ratio Level of Measurement

The king of data types is the *ratio* level. This is as good as it gets as far as data is concerned. Now we can perform all four mathematical operations to compare values with absolutely no feelings of guilt. Examples of this type of data are age, weight, height, and salary. Ratio data has all the features of interval data with the added benefit of a true 0 point. The term "true zero point" means that a 0 data value indicates the absence of the object being measured. For instance, 0 salary indicates the absence of any salary.

Wrong Number

Interval data does not have a true 0 point. For example, 0 degrees Fahrenheit does not represent the absence of temperature, even though it may feel like it. To help explain this, try baking a cake at twice the recommended temperature in half the recommended time. Yuck!

With a true 0 point, the rules of multiplication and division can be used to compare data values. This allows us to say that a person who is 6 feet in height is twice as tall as a 3-foot person or that a 20-year-old person is half the age of a 40 year old.

The distinction between interval and ratio data is a fine line. To help identify the proper scale, use the "twice as much" rule. If the phrase "twice as much" accurately describes the relationship between 2 values that differ by a multiple of 2, then the data can be considered ratio level.

There are endless examples of ratio data. Let's look at measuring typing speed in words per minute. I happen to be a handicapped, 2-finger, hunt-and-peck typist whose has tried those darned typing programs more than once and just can't get it. I can type maybe 20 words a minute on a good day. My 15-year-old son, John, on the other hand, is one of those show-offs who types while he's *not even looking* and can still type 60 words a minute. Because we can correctly say that John types 3 times faster than me, typing speed is an example of ratio data.

Figure 2.1 summarizes the different data scales and how they relate to one another. As we explore different statistical techniques later in this book, we will revisit these different measurement scales. You will discover that specific techniques require certain types of data.

Figure 2.1

Summary of data measurement scales.

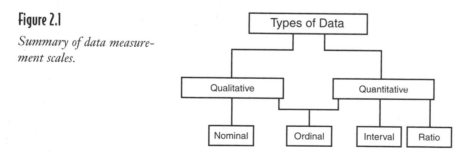

Computers to the Rescue

As mentioned in Chapter 1, we will explore the use of Excel in solving some of the statistics problems in this book. If you have no interest in using Excel in this manner, just skip over this section. I promise you won't hurt my feelings. The purpose of this last section is to talk about the use of computers with statistics in general and then to make sure your computer is ready to follow us along.

The Role of Computers in Statistics

When I was a youthful engineering undergraduate student during the 1970s, the words "personal computer" had no meaning. Calculations were performed on a clever gadget fondly known as a "slide rule." For those of you who weren't even alive during this time period, I've included a picture of one in Figure 2.2.

Figure 2.2

Slide rule circa 1975.

(www.hpmuseum.org)

As you can see for yourself, this device looks like a ruler on steroids. It can perform all sorts of mathematical calculations, but is far from being user friendly. During my freshman year in college, I purchased my first handheld calculator, a Texas Instrument model that could only perform the basic math functions. It was the approximate size of a cash register.

At this point, the only serious statistical analysis was being performed on mainframe computers by people with high levels of programming skill. These people were some-what "different" from the rest of us. Fortunately, we have advanced from the Dark Ages and now have awesome, user-friendly computing power at our fingertips. Powerful programs such as SAS, SPSS, Minitab, and Excel are readily accessible to those of us who don't know a lick of computer programming and allow us to perform some of the most sophisticated statistical analysis known to mankind.

Parts of this book will demonstrate how to solve some of the statistical techniques using Microsoft Excel. If you choose to skip over these parts, it will not interfere with your grasp of topics in subsequent chapters. This is simply optional material to expose you to statistical analysis on the computer. My assumption is going to be that you already have a basic working knowledge of how to use Excel.

Installing the Data Analysis Add-in

Our first task is to check whether Excel's data analysis tools are available on your computer. Open the Excel program and left-click with your mouse on the Tools menu as shown in Figure 2.3. From this point on in the book, I'll use the term "click" to mean click the left button on your mouse.

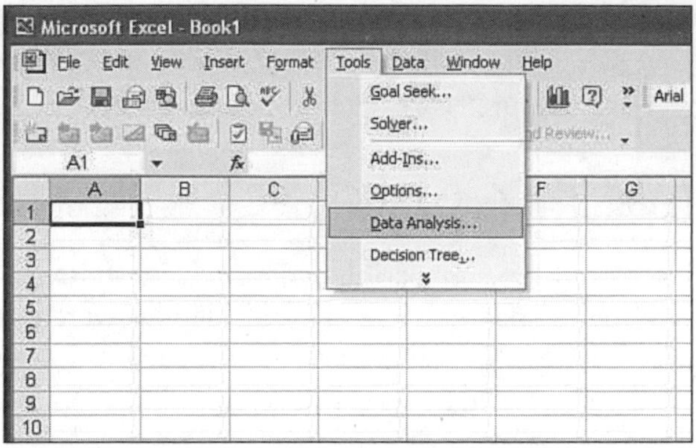

Figure 2.3

Excel's Tools menu.

Notice in the figure that the highlighted item is Data Analysis, which may or may not show up under your Tools menu. If Data Analysis does appear under your Tools menu, skip the rest of this paragraph and the next two and proceed to the following paragraph that begins with "Click on Data Analysis."

> ### Random Thoughts
>
> If your Tools menu looks different from the one in Figure 2.3, it might be because all of your available menu items are not currently visible. To make all the menu items visible, click on the Expand symbol at the bottom of the list (the double-downward-pointing arrows).

If Data Analysis does not appear under the Tools menu, you need to add it to the menu. To do so, click on Add-Ins under the same Tools menu. If you don't see Add-Ins under this menu at first, expand the menu items by clicking on the downward arrow at the bottom of the Tools menu list. After clicking on Add-Ins, you should see the dialog box in Figure 2.4.

Figure 2.4

Excel's Add-Ins dialog box.

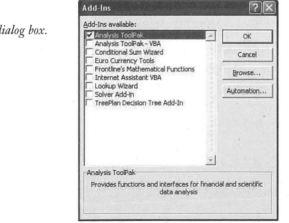

This dialog box provides a list of available add-ins for you to use. Click on the empty box for Analysis ToolPak, which places a check mark in it, and then click OK. Now click on the Tools menu again; Data Analysis will now appear in the list.

Random Thoughts

Don't panic if you receive the following message: "Microsoft Excel can't run this add-in. This feature is not currently installed. Would you like to install it now?" If you want to install the Analysis ToolPak, you might need to have your original Microsoft Office CD close by. Click the Yes button and follow any further instructions. Then, the Data Analysis option will become available on the Tools menu.

Click on Data Analysis under the Tools menu to open the dialog box shown in Figure 2.5.

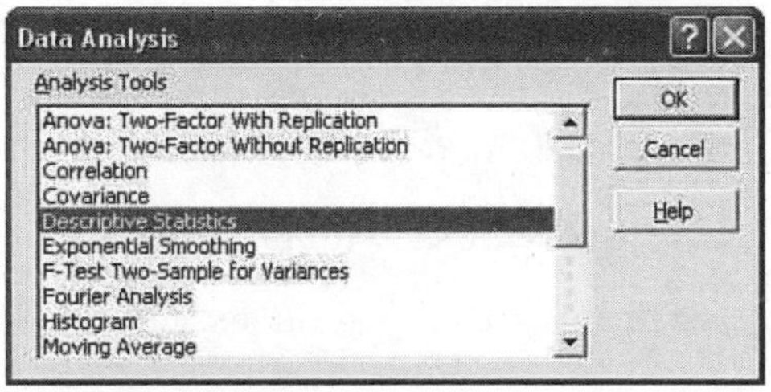

Figure 2.5

Excel's Data Analysis dialog box.

Your Excel program is now ready to perform all sorts of statistical magic for you as we explore various techniques throughout this book. At this point, you can click Cancel and close out Excel. Each time that you open Excel in the future, the Data Analysis option will be available under the Tools menu.

Your Turn

Classify the following data as nominal, ordinal, interval, or ratio. Explain your choice.

1. Average monthly temperature in degrees Fahrenheit for the city of Wilmington throughout the year

2. Average monthly rainfall in inches for the city of Wilmington throughout the year

3. Education level of survey respondents

Level	Number of Respondents
High school	168
Bachelor's degree	784
Master's degree	212

4. Marital status of survey respondents

Status	Number of Respondents
Single	28
Married	189
Divorced	62

5. Age of the respondents in the survey

6. Gender of the respondents in the survey

7. The year in which the respondent was born

8. The voting intentions of the respondents in the survey classified as Republican, Democrat, or Undecided

9. The race of the respondents in the survey classified as White, African American, Asian, or Other

10. Performance rating of employees classified as Above Expectations, Meets Expectations, or Below Expectations

11. The uniform number of each member on a sports team

12. A list of the graduating high school seniors by class rank

13. Final exam scores for my statistics class on a scale of 0 to 100

14. The state in which the respondents in a survey reside

The Least You Need to Know

- Data serves as the building blocks for all statistical analysis.

- Data can be classified as either quantitative or qualitative.

- Nominal data is assigned to categories with no mathematical comparisons between observations.

- Ordinal data has all the properties of nominal data with the additional capability of arranging the observations in order.

- Interval data has all the properties of ordinal data with the additional capability of calculating meaningful differences between the observations.

- Ratio data has all the properties of interval data with the additional capability of expressing one observation as a multiple of another.

Displaying Descriptive Statistics

In This Chapter

- ◆ How to construct a frequency distribution
- ◆ How to graph a frequency distribution with a histogram
- ◆ How to construct a stem and leaf display
- ◆ The usefulness of pie, bar, and line charts
- ◆ Using Excel's Chart Wizard to construct charts

In Chapter 2, I explained the various types of data that exists for statistical analysis. In this chapter, we will explore the different ways in which data can be presented. In its basic form, it can be very difficult to make sense of the patterns in the data because our human brains are not very efficient at processing long lists of raw numbers. We do a much better job of absorbing data when it is presented in summarized form through tables and graphs.

In the next several sections, we will examine many ways to present data so that it will be more useful to the person performing the analysis. Through these techniques, we are able to get a better overview of what the data is telling us. And believe me, there is plenty of data out there with some very interesting stories to tell. Stay tuned.

Frequency Distributions

One of the most common ways to graphically describe data is through the use of *frequency distributions*. The best way to describe a frequency distribution is to start with an example.

Stat Facts

A frequency distribution is a table that shows the number of data observations that fall into specific intervals.

Ever since I was a young boy, I have been a devoted fan of the Pittsburgh Pirates Major League Baseball team. Why I still root for these guys, I'll never know, because they have not had a winning season since 1992. Anyway, below is a table of the batting averages of individual Pirates at the end of the 2002 season. I have not attached names with these averages in order to protect their identities.

Pittsburgh Pirates Final Batting Averages for the 2002 Season

.160	.300	.077	.246	.283	.125	.175	.264	.233
.264	.264	.252	.250	.294	.244	.308	.121	.100
.234	.119	.232	.216	.206	.190	.154	.150	.298

Source: www.espn.com

It is difficult to get a grasp what a tough year these guys had by just looking at this data in this table format. Transforming this data into the frequency distribution shown here makes this fact more obvious.

Batting Average	Number of Players
.000 to .049	0
.050 to .099	1
.100 to .149	4
.150 to .199	5
.200 to .249	7
.250 to .299	8
.300 to .349	2

As you can see, a frequency distribution is simply a table that organizes the number of data values into intervals. In this example, the intervals are the batting average ranges in the first column of the table. The number of data values is the number of players who fall into each interval shown in the second column. Well, there's always next season to look forward to.

The intervals in a frequency distribution are officially known as *classes*, and the number of observations in each class is known as *class frequencies*. In the next section you will learn how to arrange these classes.

Constructing a Frequency Distribution

Some important decisions need to be made when constructing a frequency distribution. To illustrate these decisions, let's use another example, something many of us can relate to—cell phones! My son John and I are on one of those "family share plans," which means he gets all the peak minutes and I get to use my phone between the hours of 3 A.M. and 6 A.M. every other Saturday. The following table represents the number of calls each day during the month of May on John's account.

Calls per Day				
3	1	2	1	1
3	9	1	4	2
6	4	9	13	15
2	5	5	2	7
3	0	1	2	7
1	8	6	9	4

Source: A very confusing phone bill that requires a Ph.D. in metaphysical telecommunications to understand

Using this data, I have constructed the following frequency distribution.

Frequency Distribution

Calls per Day	Number of Days
0–2	12
3–5	8
6–8	5
9–11	3
12–14	1
15–17	1

When arranging these classes, the following rules were followed:

1. The classes should all be of equal size. I chose 3 data values to be in each class for this distribution. An example of a class is 0–2, which includes the number of days when 0, 1, or 2 calls were made.

2. The classes need to be *mutually exclusive*, or in other words, classes should not overlap. For instance, I wouldn't want 2 classes to be 3–5 and 5–7 because 5 calls would be in 2 different classes.

Stat Facts

Classes are considered **mutually exclusive** when observations can only fall into one class. For example, the gender classes "male" and "female" are mutually exclusive because a person cannot belong to both classes.

3. Try to have no fewer than 5 classes and no more than 15 classes. Too few or too many classes tend to hide the true characteristics of the frequency distribution.

4. Avoid open-ended classes, if possible (for instance, a highest class of 15–over).

5. All data values from the original table need to be included in a class. In other words, the classes should be exhaustive.

Too few or too many classes will obscure patterns in a frequency distribution. Consider the extreme case where there are so many classes that no class has more than 1 observation. The other extreme is where there is only 1 class and all the observations reside in that class. This would be a pretty useless frequency distribution!

(A Distant) Relative Frequency Distribution

Stat Facts

Relative frequency distributions display the percentage of observations of each class relative to the total number of observations.

Another way to display frequency data is by using the *relative frequency distribution*. Rather than display the number of observations in each class, this method calculates the percentage of observations in each class by dividing the frequency of each class by the total number of observations. I can display John's data as a relative frequency distribution, as I do in the following table.

Relative Frequency Distribution

Calls per Day	Number of Days	Percentage
0–2	12	12/30 = 0.40
3–5	8	8/30 = 0.27
6–8	5	5/30 = 0.17
9–11	3	3/30 = 0.10

Calls per Day	Number of Days	Percentage
12–14	1	1/30 = 0.03
15–17	1	1/30 = 0.03
	Total = 30	Total = 1.00

According to this distribution, John uses his phone 3 to 5 times 27 percent of the days during a month.

The total percentage in a relative frequency distribution should be 100 percent or very close (within 1 percent, because of rounding errors).

Cumulative Frequency Distribution

This "kissing cousin" of the relative frequency distribution simply totals the percentages of each class as you move down the column. (Get it? Cousin, relative? Sorry, I couldn't help myself!) This provides you with the percentage of observations that are less than or equal to the class of interest. The resulting *cumulative frequency distribution* is shown here.

Stat Facts

Cumulative frequency distributions indicate the percentage of observations that are less than or equal to the current class.

Cumulative Frequency Distribution

Calls per Day	No. of Days	Percentage	Cumulative Percentage
0–2	12	12/30 = 0.40	0.40
3–5	8	8/30 = 0.27	0.67
6–8	5	5/30 = 0.17	0.84
9–11	3	3/30 = 0.10	0.94
12–14	1	1/30 = 0.03	0.97
15–17	1	1/30 = 0.03	1.00
	Total = 30	Total = 1.00	

The value 0.67 in the fourth column is the result of adding 0.40 to 0.27. According to this table, John used his phone 8 times or less on 84 percent of the days in the month.

If the frequency distribution is designed properly, it is an excellent way to tease good information out of stubborn data. The next section deals with displaying the distribution graphically.

Graphing a Frequency Distribution—the Histogram

A *histogram* is simply a bar graph showing the number of observations in each class as the height of each bar. Figure 3.1 shows the histogram for John's phone calls. I used Excel's Chart Wizard to construct this graph. I'll demonstrate how to use Chart Wizard later in this chapter (see the section "Excel's Chart Wizard of Oz").

Stat Facts

A **histogram** is a bar graph showing the number of observations in each class as the height of each bar.

This graph gives us a good visual of John's calling habits. At least the highest class on the graph is the 0 to 2 calls per day. Things could be worse.

Figure 3.1

A histogram of John's phone calls.

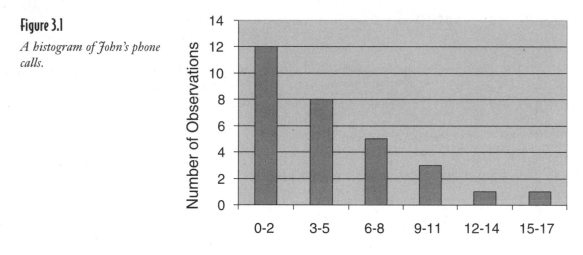

Letting Excel Do Our Dirty Work

Excel will actually construct a frequency distribution for us and plot the histogram. How nice!

1. The first thing we need to do is open Excel to a blank sheet and enter our data in Column A starting in Cell A1 (use the data from the earlier table).

Random Thoughts

For some bizarre reason, Excel refers to classes as bins. Go figure.

2. Next we enter the upper limits to each class in Column B starting in Cell B1. For example, in the class 0–2, the upper limit would be 2. Figure 3.2 shows what the spreadsheet should look like.

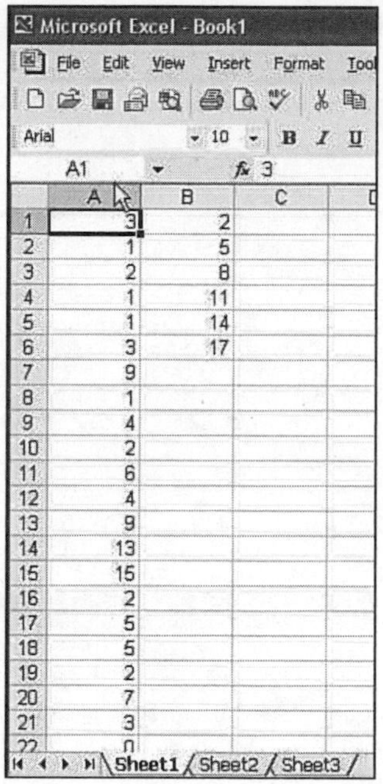

Figure 3.2

Raw data for the frequency distribution.

3. Go to the Tools menu at the top of the Excel window and select Data Analysis. (Refer to the section "Installing the Data Analysis Add-in" from Chapter 2 if you don't see the Data Analysis command on the Tools menu.)

4. Select the Histogram option from the list of Analysis Tools (see Figure 3.3) and click the OK button.

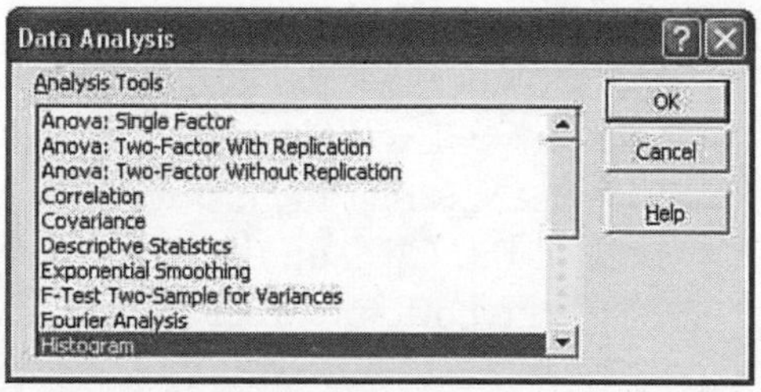

Figure 3.3

Data Analysis dialog box.

5. In the Histogram dialog box (as shown in Figure 3.4), click in the Input Range list box and then click in the worksheet to select cells A1 through A30 (the 30 original data values). Then, click in the Bin Range list box and click in the worksheet to select cells B1 through B6 (the upper limits for the 6 classes).

6. Click the New Worksheet Ply option button and the Chart Output check box (see Figure 3.4).

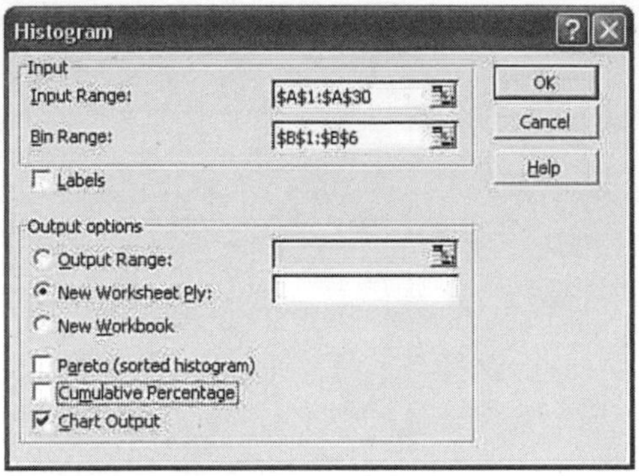

Figure 3.4

Histogram dialog box.

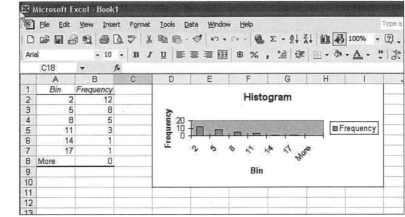

7. Click OK to generate the frequency distribution and histogram (see Figure 3.5).

Figure 3.5

Frequency distribution and histogram.

Random Thoughts

I prefer using Chart Wizard to display the histogram because I think the graph looks better than when you use the Data Analysis tool. The Chart Wizard allows you more control over the final appearance. See the section "Excel's Chart Wizard of Oz" for more information on the Chart Wizard.

Notice that the frequency distribution is generated for us by Excel in columns A and B. Cool! The problem here is that the histogram looks like an elephant sat on it. Click on the chart to select it and the click on the bottom border to drag the bottom of the chart down lower, expanding the histogram to look like Figure 3.6.

Frequency distributions and histograms are convenient ways to get an accurate picture of what your data is trying to tell you. It sounds like my data is telling me to "get more monthly minutes on your cell phone plan." Wonderful.

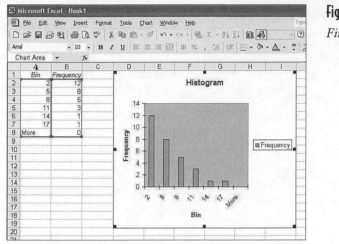

Figure 3.6

Final histogram.

Statistical Flower Power—the Stem and Leaf Display

The *stem and leaf display* is another graphical technique used to display your data. The idea was originated by a statistician named John Tukey during the 1970s. The major benefit of this approach is that all the original data points are visible on the display.

To demonstrate this method, I will use my son Brian's golf scores for his last 24 rounds, shown in the table that follows. Normally, Brian would only report his better scores, but we statisticians must be unbiased and accurate.

Brian's Golf Scores							
81	86	78	80	81	82	92	90
79	83	84	95	85	88	80	78
84	79	80	83	79	87	84	80

Figure 3.7 shows the stem and leaf display for these scores.

```
7 | 88999
8 | 00001123344445678
9 | 025

Stem and Leaf Display
```

Figure 3.7

Stem and leaf display.

The "stem" in the display is the first column of numbers, which represents the first digit of the golf scores. The "leaf" in the display is the second digit of the golf scores, with 1 digit for each score. Because there were 5 scores in the 70s, there are 5 digits to the right of 7.

Stat Facts

The **stem and leaf display** splits the data values into leaves (the last digit in the value) and stems (the remaining digits in the value). By listing all of the leaves to the right of each stem, we can graphically describe how the data is distributed.

If we choose to, we can break this display down further by adding more stems. Figure 3.8 shows this approach.

Here, the stem labeled 7 (5) stores all the scores between 75 and 79. The stem 8 (0) stores all the scores between 80 and 84. After examining this display, I can see a pattern that's not as obvious when looking at Figure 3.7: Brian usually scores in the low 80s.

An excellent source of more information about stem and leaf displays can be found at the Statistics Canada website at www.statcan.ca/english/edu/power/ch8/plots.htm.

Figure 3.8

A more detailed stem and leaf display.

```
7 (5)|88999
8 (0)|000011233444
8 (5)|5678
9 (0)|02
9 (5)|5
```
More Detailed Stem and Leaf Display

Charting Your Course

Charts are yet another efficient way to summarize and display patterns in a set of data. In this section, I will demonstrate different types of charts that help us "tell it like it is."

What's Your Favorite Pie Chart?

Pie charts are commonly used to describe data from relative frequency distributions. This type of chart is simply a circle divided into portions whose area is equal to the relative frequency distribution. To illustrate the use of pie charts, let's say some anonymous statistics professor submitted the following final grade distribution.

Final Statistics Grades

Grade	Number of Students	Relative Frequency
A	9	9/30 = 0.30
B	13	13/30 = 0.43
C	6	6/30 = 0.20
D	2	2/30 = 0.07
	Total = 30	Total = 1.00

This relative frequency distribution could be illustrated using the pie chart in Figure 3.9. This chart was done using Excel, which I will discuss later in this section.

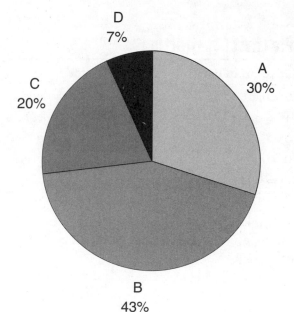

Figure 3.9

Pie chart illustrating a grade distribution.

As you can see, the pie chart approach is much easier on the eye when compared to looking at data from a table. This person must be a pretty good statistics teacher!

To construct a pie chart by hand, you first need to calculate the *center angle* for each slice in the pie, which is illustrated in Figure 3.10.

Bob's Basics

Pie charts are an excellent way to colorfully present data from a relative frequency distribution. If you cannot use colors, pie charts can also be displayed using patterns and textures.

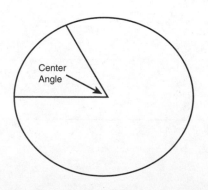

Figure 3.10

The center angle of a pie chart slice.

The center angle of each slice is determined by multiplying the relative frequency of the class by 360 (which is the number of degrees in a circle). These results are shown in the table that follows.

Center Angle for Pie Chart Construction

Grade	Relative Frequency	Central Angle
A	9/30 = 0.30	0.30 * 360 = 108 degrees
B	13/30 = 0.43	0.43 * 360 = 155 degrees
C	6/30 = 0.20	0.20 * 360 = 72 degrees
D	2/30 = 0.07	0.07 * 360 = 25 degrees
	Total = 1.00	

By using a device to measure angles, such as a protractor, you can now divide your pie chart into slices of the appropriate size. This assumes, of course, you've mastered the art of drawing circles.

Bar Charts

Bar charts are a useful graphical tool when you are plotting individual data values next to each other. To demonstrate this type of chart (see Figure 3.11), we'll use the data from the following table, which represents the monthly credit card balances for an unnamed spouse of an unnamed person writing a statistics book. (Boy, I'm going to be in *big* trouble when she sees this.)

Anonymous Credit Card Balances

Month	Balance ($)
1	375
2	514
3	834
4	603
5	882
6	468

Source: An unnamed filing cabinet

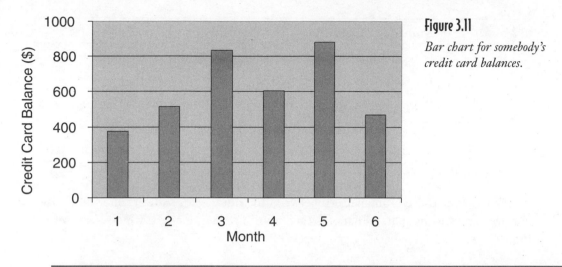

Figure 3.11

Bar chart for somebody's credit card balances.

> **Random Thoughts**
>
> By now you may have just said to yourself, "Hey, wait one minute! Haven't I seen this somewhere before?" By "this" I hope you're referring to the type of chart rather than my wife's credit card statements. The histogram that we visited earlier in the chapter is actually a special type of bar chart that plots frequencies rather than actual data values.

I'm sure your inquisitive mind is now screaming with the question "How do you choose between a pie chart and a bar chart?" If your objective is to compare the relative size of each class to one another, use a pie chart. Bar charts are more useful when you want to highlight the actual data values.

Line Charts

The last graphical tool to be discussed is a *line chart* (sometimes called a line graph), which is used to help identify patterns between two sets of data. To illustrate the use of line charts, we'll use a favorite topic of mine: teenagers and showers.

Our current resident teenagers seem to have a costly compulsion to take very long, very hot showers, and sometimes more than once a day. As I lie awake at night listening to the constant stream of hot water, all I can envision are dollar bills flowing down the drain. I have tabulated some data, which shows the number of showers the cleanest kids on the block have taken in each of the recent months with the corresponding utility bill. Notice that at these rates we average more than one shower per day.

Month	Number of Showers	Utility Bill
1	72	$225
2	91	$287
3	98	$260
4	82	$243
5	76	$254
6	85	$275

To see whether there is any pattern between the number of showers and the utility bill, we can plot the pairs of data for each month on a line chart, which is shown in Figure 3.12.

Figure 3.12

A line chart for the number of showers and the utility bill.

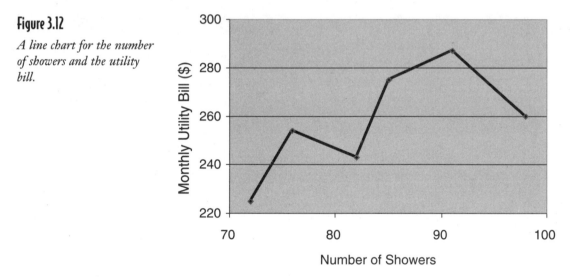

I have chosen to place the Number of Showers on the x-axis (horizontal) of the chart and the Utility Bills on the y-axis (vertical) of the chart. Because the line connecting the data points seems to have an overall upward trend, my suspicions hold true. It seems the more showers our waterlogged darlings take, the higher the utility bill.

Line charts prove very useful when you are interested in exploring patterns between 2 different types of data. They are also helpful when you have many data points and want to show all of them on 1 graph.

Excel's Chart Wizard of Oz

As promised, this last section will demonstrate how to create professional-looking charts with a few clicks of your mouse using Excel. Because pie charts are the most difficult to draw by hand, I'll show you how to create a grade distribution chart like the one that we looked at earlier in this chapter.

1. Open Excel to a blank worksheet and enter the relative frequency distribution data as shown in Figure 3.13.

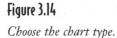

Figure 3.13

Enter the grade distribution data.

2. Start Excel's Chart Wizard—which takes you through the process step by step—by clicking on the Insert menu (between View and Format) and selecting Chart.

3. When the Chart Wizard opens, select Pie as the Chart Type and click Next at the bottom of the dialog box (see Figure 3.14).

> **Random Thoughts**
>
> You can also click the Chart Wizard button on the Standard toolbar to start Excel's Chart Wizard. The button looks like a miniature column graph. Also, if you select the cells you want to include in your chart before you start the Chart Wizard, you don't have to perform Step 4.

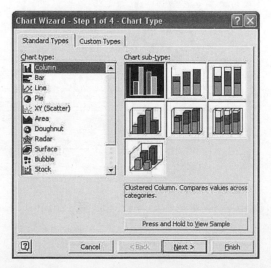

Figure 3.14

Choose the chart type.

4. If Excel doesn't automatically select the correct worksheet data for your chart, click on the Data Range text box so that your cursor appears there. Then click on cell A1 on the spreadsheet and drag your mouse over the data down to cell B5 as shown in Figure 3.15.

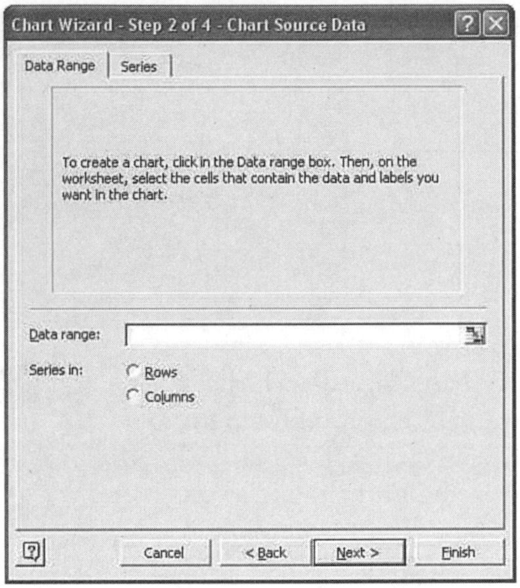

Figure 3.15

Choose the source data.

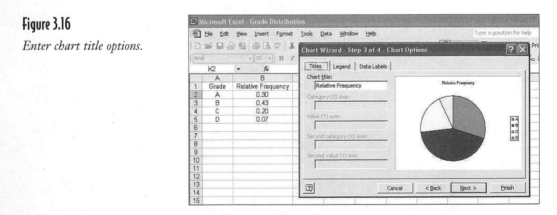

5. Be sure that the Series in: Columns option has been selected. Click Next to proceed to the chart options (see Figure 3.16).

Figure 3.16

Enter chart title options.

6. Label your pie chart as it pleases you on the Titles tab. I'll let you experiment with these on your own.

7. To add the percentages for each class on your chart, click on the Data Labels tab and check the Percentage box under the Label Contains options as shown in Figure 3.17.

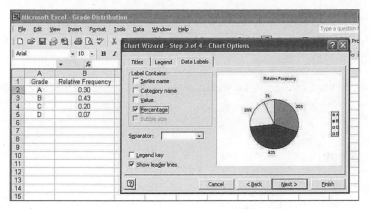

Figure 3.17

Add data label percentages to your pie chart.

8. Click Next to go back to Step 4, which gives you options for your chart location. Selecting "As new sheet" places your chart in a new sheet in the current Excel file. Selecting "As object in" places the chart into an existing sheet of your choice.

9. Click Finish and you're done! Excel creates the chart for you.

See, it's as easy as 1-2-9, and the best part is you don't have to draw a single circle.

Your Turn

1. The following table represents the exam grades from 36 students from a certain class that I might have taught. Construct a frequency distribution with 9 classes ranging from 56 to 100.

Exams Scores

60	95	75	84	85	74
81	99	89	58	66	98
99	82	62	86	85	99
79	88	98	72	72	72
75	91	86	81	96	86
78	79	83	85	92	68

2. Construct a histogram using the solution from Problem 1.

3. Construct a relative and a cumulative frequency distribution from the data in Problem 1.

4. Construct a pie chart from the solution to Problem 1.

5. Construct a stem and leaf diagram from the data in Problem 1 using 1 leaf for the scores in the 50s, 60s, 70s, 80s, and 90s.

6. Construct a stem and leaf diagram from the data in Problem 1 using 2 leaves for the scores in the 50s, 60s, 70s, 80s, and 90s.

The Least You Need to Know

◆ Frequency distributions are an efficient way to summarize data by counting the number of observations in various groupings.

◆ Histograms provide a graphical overview of data from frequency distributions.

◆ Stem and leaf displays not only provide a graphical display of the data's distribution, they also contain the actual data values of interest.

◆ Pie, bar, and line charts are effective ways to present data in different graphical forms.

◆ Excel's Chart Wizard is a powerful tool that allows the user to easily construct many different types of graphical displays.

Chapter **4**

Calculating Descriptive Statistics: Measures of Central Tendency (Mean, Median, and Mode)

In This Chapter

◆ Understanding central tendency

◆ Calculating a mean, weighted mean, median, and mode of a sample and population

◆ Calculating the mean of a frequency distribution

◆ Using Excel to calculate central tendency

The emphasis in Chapter 3 was to demonstrate ways to display our data graphically so that our brain cells could quickly absorb the big picture. With that task behind us, we can now proceed to the next step—how to summarize our data numerically. This chapter allows us to throw around some really cool words like "median" and "mode" and, when we're through, you'll actually know what they mean!

As mentioned in Chapter 1, descriptive statistics form the foundation for practically all statistical analysis. If these are not calculated with loving care, our final analysis could be misleading. And as everybody knows, statisticians never want to be misleading. So the focus of this chapter is on how to calculate descriptive statistics manually and, if you choose, to verify these results with our good friend Excel.

This is the first chapter in the book that uses mathematical formulas that have all those funny-looking Greek symbols that can make you break out into a cold sweat. Have no fear. We will slay these demons one by one through careful explanation and, in the end, victory will be ours. Onward!

Measures of Central Tendency

There exist 2 broad categories of descriptive statistics that are commonly used. The first, *measures of central tendency*, describes the center point of our data set with a single value. It's a valuable tool to help us summarize many pieces of data with 1 number. The second category, *measures of dispersion*, is the topic of Chapter 5. There are many ways to measure the central tendency of our data, and the following sections explore them.

Mean

The most common measure of central tendency is the *mean* or *average*, which is calculated by adding all the values in our data set and then dividing this result by the number of observations. The mathematical formula for the mean differs slightly depending on whether you're referring to the sample mean or the population mean. The formula for the sample mean is as follows:

$$\bar{x} = \frac{\sum_{i=1}^{n} x_i}{n}$$

where:

$\bar{x}$ = the sample mean

x_i = the values in the sample (x_1 = the first data value, x_2 = the second data value, and so on)

$\sum_{i=1}^{n} x_i$ = the sum of all the data values in the sample

n = the number of data values in the sample

Stat Facts

Measures of central tendency describe the center point of a data set with a single value. **Measures of dispersion** describe how far individual data values have strayed from the mean.

The **mean** or **average** is the most common measure of central tendency and is calculated by adding all the values in our data set and then dividing this result by the number of observations.

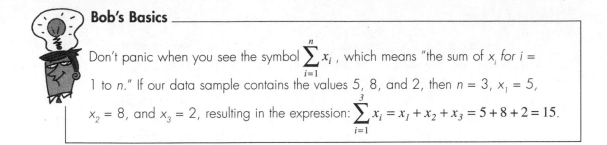

Bob's Basics

Don't panic when you see the symbol $\sum\limits_{i=1}^{n} x_i$, which means "the sum of x_i for $i = $ 1 to n." If our data sample contains the values 5, 8, and 2, then $n = 3$, $x_1 = 5$, $x_2 = 8$, and $x_3 = 2$, resulting in the expression: $\sum\limits_{i=1}^{3} x_i = x_1 + x_2 + x_3 = 5 + 8 + 2 = 15$.

The formula for the population mean is as follows:

$$\mu = \frac{\sum\limits_{i=1}^{N} x_i}{N}$$

where:

μ = the population mean (pronounced *mu*, as in "I hope you find this a*mu*sing")

$\sum\limits_{i=1}^{N} x_i$ = the sum of all the data values in the population

N = the number of data values in the population

To demonstrate calculating measures of central tendency, I'll use the following example. As in many teenage households, video games are a common form of entertainment in our family room. Brian and John love to challenge me with a game and then clean my clock before I can ask "Which team is mine?" I suspected John of sticking me with the "bad" controller because it felt like a 10-second delay between pushing a button and the game responding. Turns out the delay was really between my brain and my fingers. Anyway, here is a data set that represents the number of hours each week that video games are played in our household:

    3   7   4   9   5   4   6   17   4   7

Because this data represents a sample, we will calculate the sample mean:

$$s = \frac{\sum\limits_{i=1}^{n} x_i}{n} = \frac{3+7+4+9+5+4+6+17+4+7}{10} = 6.6 \text{ hours}$$

It looks like I need some serious practice time to catch up to these guys.

Weighted Mean

When we calculated the mean number of hours in the previous example, each data value was given the same weight in the calculation as the other values. A *weighted mean* refers to a mean that needs to go on a diet. Just kidding; I was checking to see whether you were paying attention. A *weighted mean* allows you to assign more weight to certain values and less weight to others. For example, let's say your statistics grade this semester will be based on a combination of your final exam score, a homework score, and a final project, each weighted according to the following table.

Type	Score	Weight (Percent)
Exam	94	50
Project	89	35
Homework	83	15

We can calculate your final grade using the following formula for a weighted average. Note that here we are dividing by the sum of the weights rather than by the number of data values.

$$\bar{x} = \frac{\sum_{i=1}^{n} (w_i * x_i)}{\sum_{i=1}^{n} w_i}$$

Bob's Basics

The symbol $\sum_{i=1}^{n} (w_i * x_i)$ means "the sum of w times x." Each pair of w and x is first multiplied together, and these results are then summed.

where:

w_i = the weight for each data value x_i

$\sum_{i=1}^{n} w_i$ = the sum of the weights

The previous equation can be set up in the following table to demonstrate the procedure.

Type i	Score x_i	Weight w	Weight × Score ($w * x$)
Exam 1	94	0.50	47.0
Project 2	89	0.35	31.2
Homework 3	83	0.15	12.4

$$\sum_{i=1}^{3} w_i = 1.0 \qquad\qquad \sum_{i=1}^{3} (w_i * x_i) = 90.6$$

The same result can be obtained by plugging the numbers directly in to the formula for a weighted average:

$$\bar{x} = \frac{(0.50 * 94) + (0.35 * 89) + (0.15 * 83)}{0.50 + 0.35 + 0.15}$$

$$\bar{x} = \frac{47.0 + 31.2 + 12.4}{1.00} = 90.6$$

Congratulations. You earned an A.

Bob's Basics

The weights in a weighted average do not need to add up to 1 as in the previous example. Let's say I want a weighted average of my 2 most recent golf scores, 90 and 100, and I want 90 to have twice the weight as 100 in my average (if, for example, I could claim that the course in which I shot a 90 was twice as hard as the course in which I shot 100). I would calculate this as $\bar{x} = \dfrac{(2 * 90) + (1 * 100)}{3} = 93.3$. By giving more weight to my lower score, the result is lower than the true average of 95. In this case, I think I'll go with the weighted average.

Mean of Grouped Data from a Frequency Distribution

Here is some great news to get excited about. You can actually calculate the mean of grouped data from a frequency distribution. Recall the data set from Chapter 3 regarding John's cell phone calls per day shown in the following table.

Calls per Day				
3	1	2	1	1
3	9	1	4	2
6	4	9	13	15
2	5	5	2	7
3	0	1	2	7
1	8	6	9	4

The following table shows this data as a frequency distribution with the Calls per Day as the class.

Frequency Distribution

Calls per Day	Number of Days
0–2	12
3–5	8
6–8	5
9–11	3
12–14	1
15–17	1

To calculate the mean of this distribution, we first need to determine the midpoint of each class using the following method:

$$\text{Class Midpoint} = \frac{\text{LowerValue} + \text{UpperValue}}{2}$$

For instance, the class midpoint for the last class would be as follows:

$$\text{Class Midpoint} = \frac{15 + 17}{2} = 16$$

We can use the following table to assist in the calculations.

Class	Midpoint (x)	Frequency (f)
0–2	1	12
3–5	4	8
6–8	7	5
9–11	10	3
12–14	13	1
15–17	16	1

After the midpoint for each class has been determined, we can calculate the mean of the frequency distribution using the following equation—which is basically a weighted average formula:

$$\bar{x} = \frac{\sum_{i=1}^{n} (f_i * x_i)}{\sum_{i=1}^{n} f_i}$$

where:

x_i = the midpoint for each class for i = 1 to n

f_i = the number of observations (frequency) of each class for i = 1 to n

n = the number of classes in the distribution

The mean of this frequency distribution is determined as follows:

$$\bar{x} = \frac{(12*1)+(8*4)+(5*7)+(3*10)+(1*13)+(1*16)}{12+8+5+3+1+1} = 4.6 \text{ calls}$$

According to the mean of this frequency distribution, John averages 4.6 calls per day with his cell phone.

CAUTION

Wrong Number _____

The mean of a frequency distribution where data is grouped into classes is only an approximation to the mean of the original data set from which it was derived. This is true because we make the assumption that the original data values are at the midpoint of each class, which is not necessarily the case. The true mean of the 30 original data values in the cell phone example is only 4.5 calls per day rather than 4.6.

If the classes in the frequency distribution are a single value rather than an interval, the mean is calculated by treating the distribution as a weighted mean. For example, let's say that the following table represents the number of days that a hardware store experienced various daily demands for a particular hammer during the past 65 days of business.

Daily Demand (x)	Number of Days (f)
0	10
1	15
2	12
3	18
4	6
5	4
	Total = 65

For instance, there were 15 days in the past 65 days that the store experienced demand for 1 hammer. What is the average daily demand during the past 65 days?

$$\bar{x} = \frac{\sum_{i=1}^{n}(f_i * x_i)}{\sum_{i=1}^{n} f_i}$$

$$\bar{x} = \frac{(10*0)+(15*1)+(12*2)+(18*3)+(6*4)+(4*5)}{10+15+12+18+6+4}$$

$$\bar{x} = \frac{137}{65} = 2.1 \text{ hammers per day}$$

Now that we have become experts in every conceivable method for calculating a mean, we are ready to move on to the other cool methods to measure central tendency.

Median

The mean isn't the only way to measure central tendency. The *median* is the value in the data set for which half the observations are higher and half the observations are lower. We find the median by arranging the data values in ascending order and identifying the halfway point.

Going back to our example with the video games, we see our data set rearranged in ascending order:

3   4   4   4   5   6   7   7   9   17

Because we have an even number of data points (10), the median is the average of the 2 center points. In this case that will be the values 5 and 6, resulting in a median of 5.5 hours of video games per week. Notice that there are 4 data values to the left (3, 4, 4, and 4) of these center points and 4 data values to the right (7, 7, 9, and 17).

Stat Facts

The **median** is a measure of central tendency that represents the value in the data set for which half the observations are higher and half the observations are lower. When there is an even number of data points, the median will be the average of the 2 center points.

To illustrate the median for a data set with an odd number of values, let's remove 17 from the video games data and repeat our analysis.

3   4   4   4   5   6   7   7   9

In this instance, we only have 1 center point, which is the value 5. Therefore, the median for this data set is 5 hours of video games per week. Again, there are 4 data values to the left and right of this center point.

Mode

The last measure of central tendency on my mind is the *mode*, which is simply the observation in the data set that occurs the most frequently.

To illustrate the mode for a data set, let's use the original video games data.

3   4   4   4   5   6   7   7   9   17

The mode is 4 hours per week because this value occurs 3 times in the data set.

That wraps up all the different ways to measure central tendency of our data set. However, there is 1 question screaming to be answered, and that being ....

Random Thoughts
There can be more than 1 mode of a data set if more than 1 value occurs the most frequent number of times.

How Does One Choose?

I bet you never thought you would have so many choices of measuring central tendency? It's kind of like being in an ice cream store in front of 30 flavors. If you think that all of the data in your data set is relevant, then the mean is your best choice. This measurement is affected by both the number and magnitude of your values. However, very small or very large values can have a significant impact on the mean, especially if the size of the sample is small. If this is a concern, perhaps you should consider using the median. The median is not as sensitive to a very large or small value.

Consider the following data set from the original video game example:

3   4   4   4   5   6   7   7   9   17

The number 17 is rather large when compared to the rest of the data. The mean of this sample was 6.6, whereas the median was 5.5. If you thought 17 was not a typical value that you would expect in this data set, the median would be your best choice for central tendency.

The poor lonely mode has limited applications. It is primarily used to describe data at the nominal scale—that is, data that is grouped in descriptive categories such as gender. If 60 percent of our survey respondents were male, then the mode of our data would be male.

Using Excel to Calculate Central Tendency

Excel will kindly calculate the mean, median, and mode for you all at once with a few mouse clicks. I will demonstrate this using the data set from the video game example.

1. To begin, open a blank Excel worksheet and enter the video game data as shown in Figure 4.1.

Figure 4.1

Enter data from the video game example.

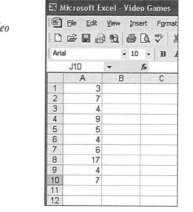

2. Click on the Tools menu at the top of the spreadsheet (between Format and Data) and select Data Analysis. (See the section "Installing the Data Analysis Add-in" in Chapter 2 for more details on this step if you don't see the Data Analysis command.) After selecting Data Analysis, you should see the following dialog box (Figure 4.2).

Figure 4.2

Data Analysis dialog box.

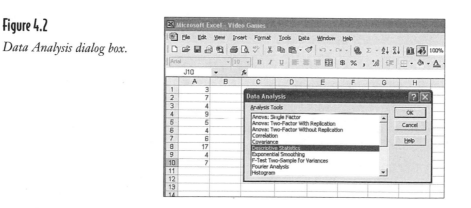

3. Select Descriptive Statistics and click OK. The following dialog box will appear (Figure 4.3).

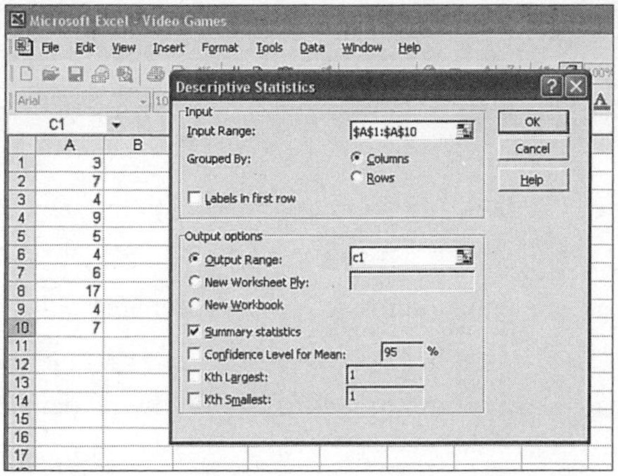

Figure 4.3

Descriptive Statistics dialog box.

4. For the Input Range, select cells A1 through A10, select the Output Range option, and select cell C1. Then choose the Summary statistics check box and click OK.

5. After you expand columns C and D slightly to see all the figures, your spreadsheet should look like Figure 4.4.

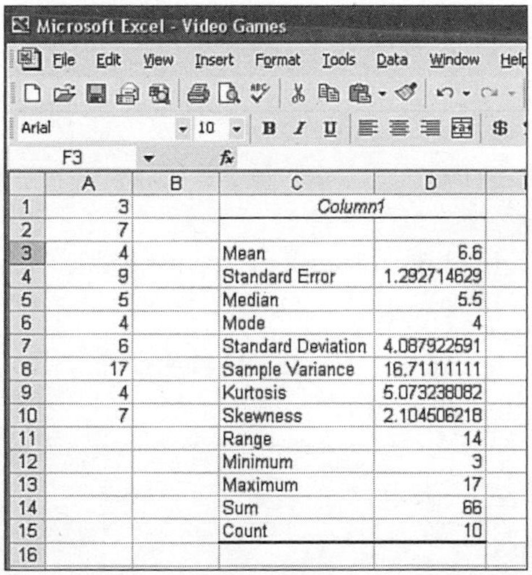

	A	B	C	D
1	3		*Column1*	
2	7			
3	4		Mean	6.6
4	9		Standard Error	1.292714629
5	5		Median	5.5
6	4		Mode	4
7	6		Standard Deviation	4.087922591
8	17		Sample Variance	16.71111111
9	4		Kurtosis	5.073238082
10	7		Skewness	2.104506218
11			Range	14
12			Minimum	3
13			Maximum	17
14			Sum	66
15			Count	10
16				

Figure 4.4

Measures of central tendency for the video game example.

As you can see, the mean is 6.6 hours, the median is 5.5 hours, and the mode is 4.0 hours. Piece of cake!

Your Turn

1. Calculate the mean, median, and mode for the following data set: 20 15 24 10 8 19 24 12 21 6

2. Calculate the mean, median, and mode for the following data set: 84 82 90 77 75 77 82 86 82

3. Calculate the mean, median, and mode for the following data set: 36 27 50 42 27 36 25 40 29 15

4. Calculate the mean, median, and mode for the following data set: 8 11 6 2 11 6 5 6 10

5. A company counted the number of their employees in each of the following age classes. According to this distribution, what is the average age of the employees in the company?

Age Range	Number of Employees
20–24	8
25–29	37
30–34	25
35–39	48
40–44	27
45–49	10

6. Calculate the weighted mean of the following values with the corresponding weights.

Value	Weight
118	3
125	2
107	1

7. A company counted the number of employees at each level of years of service in the following table. What is the average number of years of service in this company?

Years of Service	Number of Employees
1	5
2	7
3	10
4	8
5	12
6	3

The Least You Need to Know

◆ The mean of a data set is calculated by summing all the values and dividing this result by the number of values.

◆ The median of a data set is the midpoint of the set if the values were arranged in ascending or descending order.

◆ The median is the single center value from the data set if there are an odd number of values in the set. The median is the average of the 2 center values if the number of values in the set is even.

◆ The mode of a data set is the value that appears most often in the set. There can be more than 1 mode in a data set.

Chapter 5

Calculating Descriptive Statistics: Measures of Dispersion

In This Chapter

- ◆ Calculating the range of a sample
- ◆ Calculating the variance and standard deviation of a sample and population
- ◆ Using the empirical rule and Chebyshev's theorem to predict the distribution of data values
- ◆ Using measures of relative position to identify outlier data values
- ◆ Using Excel to calculate measures of dispersion

In Chapter 4, we calculated measures of central tendency by summarizing our data set into a single value. But in doing so, we lose information that could be useful. For the video game example, if the only information I provided you was that the mean of my sample was 6.6 hours, you would not know whether all the values were between 6 and 7 hours or whether the values varied between 1 and 12 hours. As you will see later, this distinction can be very important.

To address this issue, we rely on the second major category of descriptive statistics, measures of dispersion, which describe how far the individual data values have strayed from the mean. The following sections describe different ways in which dispersion can be measured.

Range

The *range* is the simplest measure of dispersion and is calculated by the difference between the highest value and the lowest value in the data set. To demonstrate how to calculate the range, I'll use the following example. One of Debbie's special qualities is that she is a dedicated "grill-a-holic" when it comes to barbecuing in the backyard. And as such, the following data set represents the number of meals each month that Deb cranks up the grill:

7 9 8 11 4

Stat Facts

The **range** of a sample is obtained by subtracting the smallest measurement from the largest measurement.

The range of this sample would be:

Range = 11 – 4 = 7 meals

The range is a "quick-and-dirty" way to get a feel for the spread of the data set. However, the limitation is that it only relies on 2 data points to describe the variation in the sample. No other values between the highest and lowest points are part of the range calculation.

Variance

One of the most common measurements of dispersion in statistics is the *variance*, which summarizes the squared deviation of each data value from the mean. The formula for the sample variance is shown here:

Stat Facts

The **variance** is a measure of dispersion that describes the relative distance between the data points in the set and the mean of the data set. This measure is widely used in inferential statistics.

$$s^2 = \frac{\sum_{i=1}^{n}(x_i - \bar{x})^2}{n-1}$$

deviation from the mean X for each data point

where:

s^2 = the variance of the sample

$\bar{x}$ = the sample mean

n = the size of the sample (the number of data values)

$(x_i - \bar{x})$ = the deviation from the mean for each value in the data set

The first step in calculating the variance is to determine the mean of the data set, which in the grilling example is 7.8 meals per month. The rest of the calculations can be facilitated by the following table.

$\bar{x}$	$\bar{x}$	$x_i - \bar{x}$	$(x_i - \bar{x})^2$
7	7.8	–0.8	0.64
9	7.8	1.2	1.44
8	7.8	0.2	0.04
11	7.8	3.2	10.24
4	7.8	–3.8	14.44

$$\sum_{i=1}^{5} (x_i - \bar{x})^2 = 26.80$$

The final sample variance calculation becomes this:

$$s^2 = \frac{26.8}{5-1} = 6.7$$

For those of us who like to do things in one step, the entire variance calculation can also be done in the following equation:

$$s^2 = \frac{(7-7.8)^2 \quad (9-7.8)^2 + (8-7.8)^2 + (11-7.8)^2 + (4-7.8)^2}{5-1} = 6.7$$

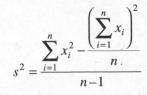

CAUTION

Wrong Number

Notice from the previous table that the result from squaring a negative number is always positive. Every number in the fourth column of this table will be a positive value.

Using the Raw Score Method (When Grilling)

A more efficient way to calculate the variance of a data set is known as the *raw score method*. Even though at first glance this equation may look more imposing, its bark is much worse than its bite. Check it out and decide for yourself what works best for you.

$$s^2 = \frac{\sum_{i=1}^{n} x_i^2 - \dfrac{\left(\sum_{i=1}^{n} x_i\right)^2}{n}}{n-1}$$

where:

$$\sum_{i=1}^{n} x_i^2 = \text{the sum of each data value after it has been squared}$$

$$\left(\sum_{i=1}^{n} x_i\right)^2 = \text{the square of the sum of all the data values}$$

Okay, don't have heart failure just yet. Let me lay this out in the following table to prove to you there are fewer calculations than the previous method.

x_i	x^2
7	49
9	81
8	64
11	121
4	16
39	331

$$\sum_{i=1}^{n} x_i = 39 \qquad \sum_{i=1}^{n} x_i^2 = 331$$

Bob's Basics

If you are calculating the variance by hand, my advice is to do your fingers and calculator battery a favor and use the raw score method.

$$\left(\sum_{i=1}^{n} x_i\right)^2 = (39)^2 = 1,521$$

$$s^2 = \frac{331 - \dfrac{1,521}{5}}{4}$$

$$s^2 = \frac{331 - 304.2}{4} = 6.7$$

As you can see, the results are the same regardless of the method used. The benefits of the raw score method become more obvious as the size of the sample (n) gets larger.

The Variance of a Population

So far, we have discussed the variance in the context of samples. The goods news is the variance of a population is calculated in the same manner as the sample variance. The bad news is I need to introduce another funny-looking Greek symbol: sigma. The equation for the variance of a population is as follows:

$$\sigma^2 = \frac{\sum_{i=1}^{N}(x_i - \mu)^2}{N}$$

where:

σ^2 = the variance of the population (pronounced "sigma squared")

x_i = the measurement of each item in the population

μ = the population mean

N = the size of the population

The raw score version of this equation is this:

$$\sigma^2 = \frac{\sum_{i=1}^{N}x_i^2 - \dfrac{\left(\sum_{i=1}^{N}x_i\right)^2}{N}}{N}$$

Wrong Number

Be sure to note that the denominator for the population variance equation is N, whereas the denominator for the sample variance is $n - 1$.

Even though this procedure is identical to the sample variance, let me demonstrate with another example. Let's say I am considering my statistics class as my population and the following ages are the measurement of interest. (Can you guess which one is me? My age adds a little spice to the variance.)

21   23   28   47   20   19   25   23

I'll use the raw score method for this calculation with the population size (N) equal to 8. (I'd love to see a class this size.)

x_i	x^2
21	441
23	529
28	784
47	2,209
20	400
19	361
25	625
23	529

$$\sum_{i=1}^{n} x_i = 206$$

$$\sum_{i=1}^{n} x_i^2 = 5,878$$

$$\left(\sum_{i=1}^{n} x_i\right)^2 = (206)^2 = 42,436$$

$$\sigma^2 = \frac{5,878 - \dfrac{42,436}{8}}{8}$$

$$\sigma^2 = \frac{5,878 - 5,304.5}{8} = 71.7$$

Thanks to the old guy in the class, the population variance is 71.7.

Standard Deviation

This one is pretty straightforward. The *standard deviation* is simply the square root of the variance. Just as with the variance, there is a standard deviation for both the sample and population, as shown in the following equations.

Stat Facts

A **standard deviation** is the square root of a variance.

Sample standard deviation:

$$s = \sqrt{s^2} = \sqrt{\frac{\sum_{i=1}^{n}\left(x_i - \overline{x}\right)^2}{n-1}}$$

Population standard deviation:

$$\sigma = \sqrt{\sigma^2} = \sqrt{\frac{\sum_{i=1}^{N}(x_i - \mu)^2}{N}}$$

To calculate the standard deviation, you must first calculate the variance and then take the square root of the result. Recall from the previous sections that the variance from my sample of the number of meals Deb grilled per month was 6.7. The standard deviation of this sample is as follows:

$$s = \sqrt{s^2} = \sqrt{6.7} = 2.6 \text{ meals}$$

Also recall the variance for the age of my class being 71.7. The standard deviation of the age of this population is as follows:

$$\sigma = \sqrt{\sigma^2} = \sqrt{71.7} = 8.5 \text{ years}$$

The standard deviation is actually a more useful measure than the variance because the standard deviation is in the units of the original data set. In comparison, the units of the variance for the grill example would be 6.7 "meals squared," and the units of the variance for the age example would 71.7 "years squared." I don't know about you, but I'm not too thrilled having my age reported as 2,209 squared years. I'll take the standard deviation over the variance any day.

Calculating the Standard Deviation of Grouped Data

The following equation shows how to calculate the standard deviation of grouped data in a frequency distribution.

$$s = \sqrt{\frac{\sum_{i=1}^{m}(x_i - \bar{x})^2 f_i}{n-1}}$$

where:

f_i = the number of data values in each frequency class

m = the number of classes

$n = \sum_{i=1}^{m} f_i$ = the total number of values in the data set

The following table is a frequency distribution that represents the number of times each child in Deb's 3-year-old preschool class needs a "potty break" in a day.

Number of Potty Breaks per Day (x_i)	Number of Children (f_i)
2	1
3	4
4	12
5	8
6	5

In this example, m = 5 and n = 30. From Chapter 4, we know the mean of this frequency distribution is this:

$$\bar{x} = \frac{\sum_{i=1}^{m}(f_i * x_i)}{\sum_{i=1}^{m} f_i}$$

$$\bar{x} = \frac{(1 \times 2) + (4 \times 3) + (12 \times 4) + (8 \times 5) + (5 \times 6)}{1 + 4 + 12 + 8 + 5} = 4.4 \text{ times per child per day}$$

The following table summarizes the standard deviation calculations.

x_i	f_i	$\bar{x}$	$\left(x_i - \bar{x}\right)$	$\left(x_i - \bar{x}\right)^2$	$\left(x_i - \bar{x}\right)^2 f_i$
2	1	4.4	−2.4	5.76	5.76
3	4	4.4	−1.4	1.96	7.84
4	12	4.4	−0.4	0.16	1.92
5	8	4.4	0.6	0.36	2.88
6	5	4.4	1.6	2.56	12.80

$$\sum_{i=1}^{m} \left(x_i - \bar{x}\right)^2 f_i = 31.20$$

$$s = \sqrt{\frac{\sum_{i=1}^{m} \left(x_i - \bar{x}\right)^2 f_i}{n - 1}} = \sqrt{\frac{31.20}{30 - 1}} = \sqrt{1.08} = 1.04 \text{ times per child per day}$$

The potty break frequency distribution has a mean of 4.4 times per child per day and a standard deviation of 1.04 times per child per day. The frequency of these potty breaks must keep Deb very busy.

The Empirical Rule: Working the Standard Deviation

The values of many large data sets tend to cluster around the mean or median so that the data distribution in the histogram resembles a bell-shape, symmetrical curve. When this is the case, the *empirical rule* (sounds like a decree from the emperor) tells us that approximately 68 percent of the data values will be within 1 standard deviation from the mean.

Stat Facts

According to the **empirical rule,** if a distribution follows a bell-shape, symmetrical curve centered around the mean, we would expect approximately 68, 95, and 99.7 percent of the values to fall within 1, 2, and 3 standard deviations around the mean respectively.

For example, suppose that the average exam score for my large statistics class is 88 points and the standard deviation is 4.0 points and that the distribution of grades is bell-shape around the mean, as shown in Figure 5.1. Because 1 standard deviation above the mean would be 92 (88 + 4) and 1 standard deviation below the mean would be 84 (88 − 4), the empirical rule tells me that approximately 68 percent of the exam scores will fall between 84 and 92 points.

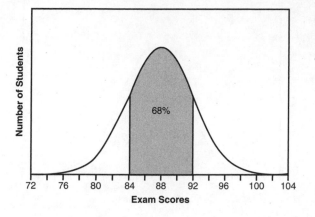

Figure 5.1

One standard deviation around the mean.

The empirical rule also states that approximately 95 percent of the data values will fall within *2* standard deviations from the mean. In our example, 2 standard deviations equal 8.0 points (2 * 40). Two standard deviations above the mean would be a score of 96 (88 + 8), and 2 standard deviations below the mean would be 80 (88 – 8). According to Figure 5.2, approximately 95 percent of the exam scores will be between 80 and 96 points.

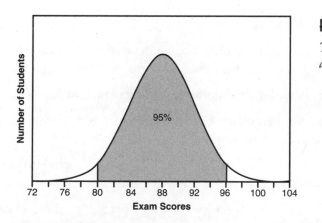

Figure 5.2

Two standard deviations around the mean.

Taking this one final step, the empirical rule states that, under these conditions, approximately 99.7 percent of the data values will fall within 3 standard deviations from the mean. According to Figure 5.3, virtually all the test scores should fall within plus or minus 12 points (3 * 40) from the mean of 88. In this case, I would expect all the exam scores to be between 76 and 100.

Figure 5.3

Three standard deviations around the mean.

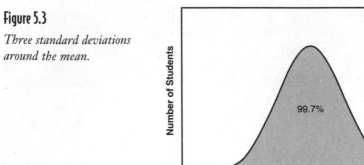

In general, we can use the following equation to express the range of values within *k* standard deviations around the mean:

$\mu - population\ mean$
$\sigma - variance$

$$\mu \pm k\sigma$$

We will revisit the empirical rule concept in subsequent chapters.

Chebyshev's Theorem

Chebyshev's theorem is a mathematical rule similar to the empirical rule except that it applies to any distribution rather than just bell-shape, symmetrical distributions. Chebyshev's theorem states that for any number *k* greater than 1, at least $\left(1 - \frac{1}{k^2}\right) * 100\%$ percent of the values will fall within *k* standard deviations from the mean. Using this equation, we can state the following:

- At least 75 percent of the data values will fall within 2 standard deviations from the mean by setting *k* = 2 into Chebyshev's equation.

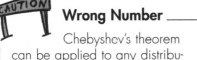

Wrong Number

Chebyshev's theorem can be applied to any distribution of data but can only be stated for values of *k* that are greater than 1.

- At least 88.9 percent of the data values will fall within 3 standard deviations from the mean by setting *k* = 3 into the equation.

- At least 93.7 percent of the data values will fall within 4 standard deviations from the mean by setting *k* = 4 into the equation. This last example is shown as

$$\left(1 - \frac{1}{4^2}\right) * 100\% = 93.7\% \,.$$

Let's check out Chebyshev's theorem to see whether it really works. The following table shows the number of home runs hit by the top 40 players in Major League Baseball during the 2002 season.

Number of Home Runs from Top 40 Players in 2002									
57	52	49	46	43	42	42	41	39	39
38	38	37	37	35	34	34	34	33	33
33	32	31	31	31	30	30	30	29	29
29	29	29	28	28	28	28	28	27	27

Source: www.espn.com.

The following histogram shows that this distribution is neither bell-shape nor symmetrical, so the empirical rule cannot be applied (see Figure 5.4). We will need to use Chebyshev's theorem.

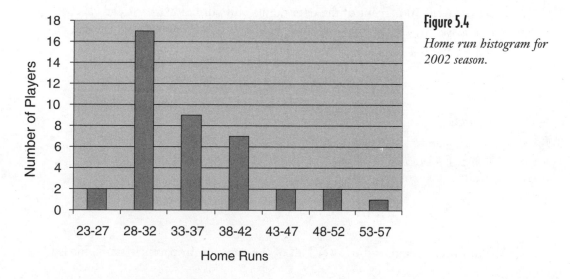

Figure 5.4

Home run histogram for 2002 season.

The mean for this distribution is 34.7 home runs, and the standard deviation is 7.2 home runs. The table that follows summarizes various intervals around the mean with the percentage of values within those intervals.

k	μ	σ	$\mu + k\sigma$	$\mu - k\sigma$	Chebyshev's %	Actual %
2	34.7	7.2	49.1	20.3	75.0%	95.0%
3	34.7	7.2	56.3	13.1	88.9%	97.5%
4	34.7	7.2	63.5	5.9	93.7%	100.0%

This table supports Chebyshev's theorem, which predicts that at least 75 percent of the values will fall within 2 standard deviations from the mean. From the data set, we can observe that 95 percent actually fall between 20.3 and 49.1 home runs (38 out of 40). The same explanation holds true for 3 and 4 standard deviations around the mean.

Measures of Relative Position

Another way of looking at dispersion of data is through measures of relative position, which describe the percentage of the data below a certain point. The following sections describe this technique.

Quartiles

Quartiles divide the data set into 4 equal segments after it has been arranged in ascending order. Approximately 25 percent of the data points will fall below the *first quartile*, Q_1. Approximately 50 percent of the data points will fall below the *second quartile*, Q_2. And, you guessed it, 75 percent should fall below the *third quartile*, Q_3. To demonstrate how to identify Q_1, Q_2, and Q_3, let's use the following data set.

9   5   3   10   14   6   12   7   14

Stat Facts

Quartiles measure the relative position of the data values by dividing the data set into 4 equal segments.

Step 1: Arrange your data in ascending order.

3   5   6   7   (9)   10   12   14   14

Step 2: Find the median of the data set. This is Q_2.

3   5   6   7   9   10   12   14   14

$Q_2 = 9$

Step 3: Find the median of the lower half of the data set (in parenthesis). This is Q_1.

(3   5   6   7)   9   10   12   14   14

$Q_1 = 5.5$

$Q_2 = 9$

Step 4: Find the median of the upper half of the data set (in parenthesis). This is Q_3.

3   5   6   7   9   (10   12   14   14)

$Q_1 = 5.5$

$Q_2 = 9$

$Q_3 = 13$

Interquartile Range

When the quartiles are established, the *interquartile range* (IQR) can be easily calculated; the IQR measures the spread of the center half of our data set. It is simply the difference between the third and first quartiles, as follows:

$$IQR = Q_3 - Q_1$$

$$IQR = 13 - 5.5 = 7.5$$

The interquartile range is used to identify outliers, which are the "black sheep" of our data set. These are extreme values whose accuracy is questioned and can cause unwanted distortions in statistical results. Any values that are more than:

Stat Facts

The **interquartile range** measures the spread of the center half of the data set and is used to identify outliers, which are extreme values that should be discarded before analysis.

$$Q_3 + 1.5IQR$$

$$13 + 1.5(7.5) = 24.25$$

or less than:

$$Q_1 - 1.5IQR$$

$$5.5 + 1.5(7.5) = -5.75$$

should be discarded.

Now that we have worked our fingers to the bones calculating all this stuff, let's see how Excel makes it look so easy.

Using Excel to Calculate Measures of Dispersion

Excel enables you to conveniently calculate the range, variance, and standard deviation of a sample using the Data Analysis selection under the Tools menu. The steps needed to calculate these measures are the exact same as those used to calculate measures of central tendency shown in Chapter 4. Repeating those steps (see the section "Using Excel to Calculate Central Tendency") with the grilling example from this chapter will produce Figure 5.5.

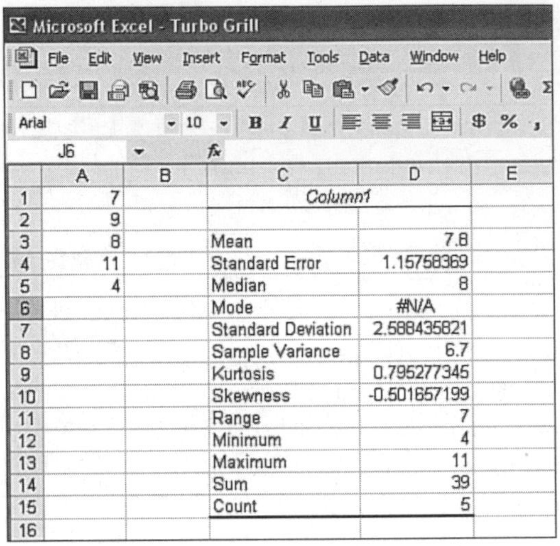

Microsoft Excel - Turbo Grill

	A	B	C	D	E
1	7		Column1		
2	9				
3	8		Mean	7.8	
4	11		Standard Error	1.15758369	
5	4		Median	8	
6			Mode	#N/A	
7			Standard Deviation	2.588435821	
8			Sample Variance	6.7	
9			Kurtosis	0.795277345	
10			Skewness	-0.501657199	
11			Range	7	
12			Minimum	4	
13			Maximum	11	
14			Sum	39	
15			Count	5	
16					

Figure 5.5
Measures of dispersion for the turbo grill example.

> CAUTION
> **Wrong Number**
> The values for variance and standard deviation reported by Excel are for a sample. If your data set represents a population, you need to recalculate the results using N in the denominator rather than $n - 1$.

As you can see from Figure 5.5, the sample range equals 7 meals, the sample variance equals 6.7, and the standard deviation equals 2.6 meals. Also note that this data set has no mode since no value appears more than once.

This wraps up our discussion on the different ways to describe measures of dispersion.

Your Turn

1. Calculate the variance, standard deviation, and the range for the following sample data set: 20 15 24 10 8 19 24
2. Calculate the variance, standard deviation, and the range for the following population data set: 84 82 90 77 75 77 82 86 82
3. Calculate the variance, standard deviation, and the range for the following sample data set: 36 27 50 42 27 36 25 40
4. Calculate the quartiles and the cutoffs for the outliers for the following data set: 8 11 6 2 11 6 5 6 10 15
5. A company counted the number of their employees in each of the age classes as follows. According to this distribution, what is the standard deviation for the age of the employees in the company?

Age Range	Number of Employees
20–24	8
25–29	37
30–34	25
35–39	48
40–44	27
45–49	10

6. A company counted the number of employees at each level of years of service in the table that follows. What is the standard deviation for the number of years of service in this company?

Years of Service	Number of Employees
1	5
2	7
3	10
4	8
5	12
6	3

7. A data set that follows a bell-shape and symmetrical distribution has a mean equal to 75 and a standard deviation equal to 10. What range of values centered around the mean would represent 95 percent of the data points?

8. A data set that is not bell-shape and symmetrical has a mean equal to 50 and a standard deviation equal to 6. What is the minimum percent of values that would fall between 38 and 62?

The Least You Need to Know

- The range of a data set is the difference between the largest value and smallest value.

- The variance of a data set summarizes the squared deviation of each data value from the mean.

- The standard deviation of a data set is the square root of the variance and is expressed in the same units as the original data values.

◆ The empirical rule states that if a distribution follows a bell-shape, symmetrical curve centered around the mean, we would expect approximately 68, 95, and 99.7 percent of the values to fall within 1, 2, and 3 standard deviations around the mean, respectively.

◆ The interquartile range measures the spread of the center half of the data set and is used to identify outliers, which are extreme values that should be discarded before analysis.

Part 2

Probability Topics

The connection between descriptive and inferential statistics is based on probability concepts. I know the topic of probability theory scares the living daylights out of many students, but it is a very important topic in the world of statistics. The topic of probability acts as a critical link between descriptive and inferential statistics. Without a firm grasp of probability concepts, inferential statistics will seem like a foreign language. Because of this, Part 2 is designed to help you through this hurdle.

Introduction to Probability

In This Chapter

♦ Distinguishing between classical, empirical, and subjective probability

♦ Using frequency distributions to calculate probability

♦ Examine the basic properties of probability

♦ Demonstrate the intersection and union of simple events using a Venn diagram

As we leave the happy world of descriptive statistics, you may feel like you're ready to take on the challenge of inferential statistics. But before we enter that realm, we need to arm ourselves with probability theory. Accurately predicting the probability that an event will occur has widespread applications. For instance, the gaming industry uses probability theory to set odds for lotteries, card games, and sporting events.

The focus of this chapter is to start with the basics of probability, after which we will gently proceed to more complex concepts in Chapters 7 and 8. We'll discuss different types of probabilities and how to calculate the probability of simple events. We'll rely on data from frequency distributions to examine the likelihood of a combination of simple events. So pull up a chair and let's roll those dice!

What Is Probability?

Most of our daily lives are surrounded by probability concepts. When I see the weather forecast showing an 80 percent chance of rain tomorrow when I want to play golf or that my beloved Pittsburgh Pirates have only won 40 percent of their games this year (which they also did *last* year and the *year before that*), there is a 65 percent chance I will get moody.

In simple terms, probability is the likelihood of a particular event like rain or winning a ballgame. But before we go any further, we need to tackle some new "stat jargon." The following terms are widely used when talking about probability:

◆ **Experiment.** The process of measuring or observing an activity for the purpose of collecting data. An example is rolling a pair of dice.

◆ **Outcome.** A particular result of an experiment. An example is rolling a pair of 3s with the dice.

◆ **Sample space.** All the possible outcomes of the experiment. The sample space for our experiment is the numbers {2, 3, 4, 5, 6, 7, 8, 9, 10, 11, and 12}. Statistics people like to put {} around the sample space values because they think it looks cool.

◆ **Event.** One or more outcomes that are of interest for the experiment and which is/are a subset of the sample space. An example is rolling a total of 2, 3, 4, or 5 with the 2 dice.

To properly define probability, we need to consider which type of probability we are referring to.

Classical Probability

Classical probability refers to situations when we know the number of possible outcomes of the event of interest and can calculate the probability of that event with the following equation:

$$P[A] = \frac{\text{Number of possible outcomes in which Event A occurs}}{\text{Total number of possible outcomes in the sample space}}$$

where:

$P[A]$ = the probability that Event A will occur

For example, if Event A = rolling a total of 2, 3, 4, or 5 with 2 dice, we need to define the sample space for this experiment, which is shown in the following table.

{1,1}	{2,1}	{3,1}	{4,1}	{5,1}	{6,1}
{1,2}	{2,2}	{3,2}	{4,2}	{5,2}	{6,2}
{1,3}	{2,3}	{3,3}	{4,3}	{5,3}	{6,3}
{1,4}	{2,4}	{3,4}	{4,4}	{5,4}	{6,4}
{1,5}	{2,5}	{3,5}	{4,5}	{5,5}	{6,5}
{1,6}	{2,6}	{3,6}	{4,6}	{5,6}	{6,6}

There are 36 total outcomes for this experiment, each with the same chance of occurring. The outcomes that correspond to Event A are underlined. There is a total of 10 of them. Therefore:

$$P[A] = \frac{10}{36} = 0.28$$

To use classical probability, you need to understand the underlying process so that you can determine the number of outcomes associated with the event. You also need to be able to count the total number of possible outcomes in the sample space. As you will see next, this may not always be possible.

Stat Facts

Classical probability requires that you know the number of outcomes that pertain to a particular event of interest. You also need to know the total number of possible outcomes in the sample space.

Empirical Probability

When we don't know enough about the underlying process to determine the number of outcomes associated with an event, we rely on *empirical probability*. This type of probability observes the number of occurrences of an event through an experiment and calculates the probability from a relative frequency distribution. Therefore:

$$P[A] = \frac{\text{Frequency in which Event A occurs}}{\text{Total number of observations}}$$

One example of empirical probability is to answer the age-old question "What is the probability that John will get out of bed in the morning for school after his first wake-up call?" Because I cannot begin to understand the underlying process of why a teenager will resist getting out of bed before 2 P.M., I need to rely on empirical probability. The following table indicates the number of wake-up calls John required over the past 20 school days.

Stat Facts

Empirical probability requires that you count the frequency that an event occurs through an experiment and calculate the probability from the relative frequency distribution.

John's Wake-Up Calls (Previous 20 School Days)									
2	4	3	3	1	2	4	3	3	1
4	2	3	3	1	3	2	4	3	4

We can summarize this data with a relative frequency distribution.

Relative Frequency Distribution for John's Wake-Up Calls

Number of Wake-Up Calls	Number of Observations	Percentage
1	3	3/20 = 0.15
2	4	4/20 = 0.20
3	8	8/20 = 0.40
4	5	5/20 = 0.25
	Total = 20	

Based on these observations, if Event A = John getting out of bed on the first wake-up call, then P[A] = 0.15.

Using the previous table, we can also examine the probability of other events. Let's say Event B = John requiring more than 2 wake-up calls to get out of bed; then P[B] = 0.40 + 0.25 = 0.65. That boy needs to go to bed earlier on school nights!

If I choose to run another 20-day experiment of John's waking behavior, I would most likely see different results than those in the previous table. However, if I were to observe 100 days of this data, the relative frequencies would approach the true or classical probabilities of the underlying process. This pattern is known as the *law of large numbers*.

Stat Facts

The **law of large numbers** states that when an experiment is conducted a large number of times, the empirical probabilities of the process will converge to the classical probabilities.

To demonstrate the law of large numbers, let's say I flip a coin 3 times and each time the result is heads. For this experiment, the empirical probability for the event heads is 100 percent. However, if I were to flip the coin 100 times, I would expect the empirical probability of this experiment to be much closer to the classical probability of 50 percent.

Subjective Probability

Subjective probability is used when classical and empirical probabilities are not available. Under these circumstances, we rely on experience and intuition to estimate the probabilities.

Examples where subjective probability would be applied are "What is the probability that my son Brian will ask to borrow my new car, which happens to have a 6-speed manual transmission, for his Junior Prom?" (97 percent) or "What is the probability that my new car will come back with all 6 gears in proper working order?" (18 percent). These probabilities are based on my personal observations after returning from a "practice run" where I heard noises from my poor transmission that chilled me to the bone and to this day haunt me in my sleep. I need to use subjective probability in this situation because my car would never survive several of these "experiments."

Basic Properties of Probability

Our next step is to review the "rules and regulations" that govern probability theory. The basic ones are as follows:

♦ If P[A] = 1, then Event A must occur with certainty. An example is Event A = Deb buying a pair of shoes this month.

♦ If P[A] = 0, then Event A will not occur with certainty. An example is Event A = Bob will eventually finish the basement project that he started 3 years ago.

♦ The probability of Event A must be between 0 and 1.

♦ The sum of all the probabilities for the events in the sample space must be equal to 1. For example, if the experiment is flipping a coin with Event A = heads and Event B = tails, then A and B represent the entire sample space. We also know that P[A] + P[B] = 0.5 + 0.5 = 1.

♦ The *complement* to Event A is defined as all the outcomes in the sample space that are not part of Event A and is denoted as A'. Using this definition, we can state the following: P[A] + P[A'] = 1 or P[A] = 1 – P[A'].

For example, if the experiment is rolling a single 6-sided die, the sample space is shown in Figure 6.1.

Figure 6.1

Sample space for a single die experiment.

If we say that Event A = rolling a 1, then Event A' = rolling a 2, 3, 4, 5, or 6. Therefore:

$$P[A] = \frac{1}{6} = 0.167$$
$$P[A'] = 1 - 0.167 = 0.833$$

Up to this point, all the examples in this chapter would be considered cases of simple probability, which is defined as the probability of a single event. The following sections expand this concept to more than one event.

The Intersection of Events

Sometimes we are interested in the probability of a combination of events rather than just a simple event. To demonstrate this technique, the following table represents a frequency distribution for the grades of the 50 students from my brilliant statistics class.

Grade	Number of Students
A	18
B	22
C	10
	Total = 50

The following table, called a *contingency table*, breaks the grade distribution down by gender of the student.

Contingency Table for Grade Distribution

Grade	Male	Female	Totals
A	8	10	18
B	14	8	22
C	6	4	10
Total	28	22	50

Contingency tables show the actual or relative frequency of 2 types of data at the same time. In this case, the data types are grade and gender.

Let's say a student is chosen randomly from the class. By randomly, I mean that every student in the class has an equal chance of being chosen. We'll need to make the assumption that I don't have a "teacher's pet" in the room. We'll define Events A and B as follows:

♦ Event A = the selected student received an A grade

♦ Event B = the selected student was a female

We can use the previous table to calculate the simple probability that the selected student received an A as follows:

$$P[A] = \frac{18}{50} = 0.36$$

The probability that the selected student was a female would be as follows:

$$P[B] = \frac{22}{50} = 0.44$$

What about the probability that the student is a female *and* received an A? This event is known as the *intersection* of Events A and B and is described by $A \cap B$. The number of students from our contingency table that are female and received an A is 10, so:

$$P[A \text{ and } B] = P[A \cap B] = \frac{10}{50} = 0.20$$

The probability of the intersection of 2 events is known as a *joint probability*.

The intersection of Events A and B can also be described using Figure 6.2, known as a *Venn diagram*.

Stat Facts

A **contingency table** indicates the number of observations that are classified according to two variables. The **intersection** of Events A and B represents the number of instances where Events A and B occur at the same time (that is, the same student is both female and received an A). The probability of the intersection of 2 events is known as a **joint probability**. A **Venn diagram** is two or more overlapping circles that represent the relationship between multiple events.

The circle labeled Event A represents the 18 students who received an A grade, whereas the circle labeled Event B represents the 22 female students. The shaded area where the circles overlap represents the 10 students who are both female and received an A grade.

Figure 6.2

Shaded area is the intersection of Events A and B.

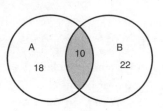

The Union of Events: A Marriage Made in Heaven

The *union* of Events A and B represents all the instances where either Event A or Event B or both occur and is denoted as $A \cup B$. Using our previous example, the following table shows the 4 groups of students who are either female or who received an A grade.

Grade	Gender	Number of Students
A	Male	8
A	Female	10
B	Female	8
C	Female	4
		Total = 30

Stat Facts

The **union** of Events A and B represents the number of instances where either Event A or B occur (that is, the number of students who are either female or who received an A).

Bob's Basics

The probability of the intersection of 2 events can never be more than the probability of the union of 2 events. If your calculations don't agree with this, go back and check for a mistake!

Therefore, the probability that the selected student is either female or received an A grade is as follows:

$$P[A \text{ or } B] = P[A \cup B] = \frac{30}{50} = 0.60$$

The union of Events A and B can also be described using a Venn diagram, as shown in Figure 6.3.

Both the middle areas represent either students who received an A grade (Circle A) or the female students (Circle B). You will learn how to calculate the probabilities of unions and intersections in Chapter 7.

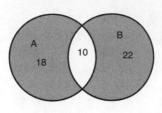

Figure 6.3

Shaded areas are the union of Events A and B.

Your Turn

1. Define each of the following as classical, empirical, or subjective probability.

    a. The probability that the baseball player Sammy Sosa will get a hit during his next at bat.

    b. The probability of drawing an Ace from a deck of cards.

    c. The probability that I will shoot lower than a 90 during my next round of golf.

    d. The probability of winning the next state lottery drawing.

    e. The probability that the drive belt for my riding lawnmower will break this summer (it did).

    f. The probability that I will finish writing this book before my deadline.

2. Identify whether each of the following are valid probabilities.

    a. 65 percent

    b. 1.9

    c. 110 percent

    d. −4.2

    e. 0.75

    f. 0

3. A survey of 125 families asked whether the household had Internet access. Each family was classified by race. The contingency table is shown here.

Race	Internet	No Internet	Total
Caucasian	15	22	37
Asian American	23	18	41
African American	14	33	47
Total	52	73	125

4. A family from the survey is randomly selected. We define:

Event A: The selected family has an Internet connection in its home.
Event B: The selected family is Asian American.

   a. Determine the probability that the selected family has an Internet connection.

   b. Determine the probability that the selected family is Asian American.

   c. Determine the probability that the selected family has an Internet connection and is Asian American.

   d. Determine the probability that the selected family has an Internet connection or is Asian American.

The Least You Need to Know

   ◆ Classical probability requires knowledge of the underlying process in order to count the number of possible outcomes of the event of interest.

   ◆ Empirical probability relies on historical data from a frequency distribution to calculate the likelihood that an event will occur.

   ◆ The law of large numbers states that when an experiment is conducted a large number of times, the empirical probabilities of the process will converge to the classical probabilities.

   ◆ The intersection of Events A and B represents the number of instances where Events A and B occur at the same time.

   ◆ The union of Events A and B represents the number of instances where either Event A or B occur.

Chapter 7

More Probability Stuff

In This Chapter

- ◆ Calculating conditional probabilities
- ◆ The distinction between independent and dependent events
- ◆ Using the multiplication rule of probability
- ◆ Defining mutually exclusive events
- ◆ Using the addition rule of probability
- ◆ Using the Bayes' theorem to calculate conditional probabilities

Now that we have arrived at the second of 3 basic probability chapters, we're ready for some new challenges. We need to take the probability concepts that you've already mastered from Chapter 6 and put them to work on the next step up the ladder. Don't worry if you're afraid of heights like I am—just keep looking up!

This chapter deals with the topic of manipulating the probability of different events in various ways. As new information about events becomes available, we can revise the old information and make it more useful. This revised information can sometimes lead to surprising results—as you'll see in this chapter.

Conditional Probability

Conditional probability is defined as the probability of Event A knowing that Event B has already occurred. To demonstrate this concept, consider this next example.

Deb is an avid tennis player and we enjoy playing matches against each other. We do, however, have one difference of opinion on the court. Deb likes to have a nice long warm-up session at the start where we hit the ball back and forth and back and forth and back and forth. All during this time, there is a little voice in my head saying, "Who's winning?" and "What's the score?" My ideal warm-up is to bend at the waist to tie my sneakers and to adjust my shorts. Each tennis match becomes a test of my manhood and the "warm-up" has nothing to do with "the thrill of victory and the agony of defeat." I can't help it; it must be a guy thing that has been passed down through thousands of years of conditioning. Deb tells me that when we rush through the warm-up, she doesn't play as well. Poppycock, I say and I'll prove it. The following table shows the outcomes of our last 20 matches, along with the type of warm-up before we started keeping score.

Contingency Table for the Tennis Example

Warm-Up Time	Deb Wins (A)	Bob Wins (A')	Total
Less than 10 min (B)	4	9	13
10 min or more (B')	5	2	7
Total	9	11	20

The events of interest are …

- ◆ Event A = Deb wins the tennis match.
- ◆ Event B = the warm-up time is less than 10 minutes.
- ◆ Event A' = Bob wins the tennis match.
- ◆ Event B' = the warm-up time is 10 minutes or more.

Without any additional information, the simple probability of each of these events is as follows:

$$P[A] = \frac{9}{20} = 0.45 \qquad P[B] = \frac{13}{20} = 0.65$$

$$P[A'] = \frac{11}{20} = 0.55 \qquad P[B'] = \frac{7}{20} = 0.35$$

As if these probabilities don't have enough names already, I have one more for you. These are also known as *prior* probabilities because they are derived only from information that is currently available.

You might ask yourself, "What other information is he talking about?" Well, suppose I know that we had a warm-up period of less than 10 minutes. Knowing this piece of info, what is the probability that Deb will win the match? This is the conditional probability of Event A given that Event B has occurred. Looking at the previous table, we can see that Event B has occurred 13 times. Because Deb has won 4 of those matches (A), the probability of A given B is calculated as follows:

Stat Facts

Simple or **prior** probabilities are always based on the total number of observations. In the previous example, that is 20 matches.

$$P[A \, / \, B] = \frac{4}{13} = 0.31$$

Deb won't be happy to see that probability.

We can also calculate the probability that Deb will win, given that the warm-up is 10 minutes or longer (otherwise known as eternity). According to the previous table, these marathon warm-ups occurred 7 times, with Deb winning 5 of these matches. Therefore:

$$P[A \, / \, B'] = \frac{5}{7} = 0.71$$

This one looks bad for Bob. I might have to hide this chapter from my live-in proof-reader.

Once again, I bring to you more "stat jargon." Conditional probabilities are also known as *posterior probabilities* (I'll resist using a butt joke here), which are considered revisions of prior probabilities using additional information. For example, the prior probability of Deb winning is P[A] = 0.45. However, with the additional information that the warm-up was 10 minutes or longer, we revise the probability of Deb winning to P[A / B'] = 0.71.

Stat Facts

Conditional probability is defined as the probability of Event A knowing that Event B has already occurred. Conditional probabilities are also known as **posterior probabilities**.

Conditional probabilities are very useful for determining the probabilities of compound events, as you will see in the following sections.

Independent Versus Dependent Events

Events A and B are said to be *independent* of each other if the occurrence of Event B has no effect on the probability of Event A. Using conditional probability, Events A and B are independent of one another if:

$$P[A / B] = P[A]$$

If Events A and B are not independent of one another, then they are said to be *dependent* events.

In the tennis example, Events A and B are dependent because the probability of Deb winning depends on whether the warm-up is more or less than 10 minutes. We can also demonstrate this by observing that:

$$P[A] = \frac{9}{20} = 0.45 \text{ and } P[A / B] = \frac{4}{13} = 0.31$$

Stat Facts

Events A and B are said to be **independent** of each other if the occurrence of Event B has no effect on the probability of Event A. If Events A and B are not independent of one another, then they are said to be **dependent** events.

These probabilities tell us that overall, Deb wins 45 percent of the matches. However, when there is a short warm-up, she only wins 31 percent of the time. Because these probabilities are not equal, Events A and B are dependent.

An example of 2 independent events is the outcome of rolling 2 dice:

♦ Event A: Roll the number 4 on the first of 2 die.

♦ Event B: Roll the number 6 on the second of 2 die.

For these events, the simple probabilities are as follows:

$$P[A] = \frac{1}{6} = 0.167 \text{ and } P[B] = \frac{1}{6} = 0.167$$

Even if we know that the first die rolled a 4, the probability of the second die being a 6 is not affected because dice, for the most part, are pretty dim-witted and are not very aware of what is going on around them. Knowing this, we can say the following:

$$P[B/A] = P[B] = \frac{1}{6} = 0.167$$

Therefore, Events A and B are independent of one another.

Multiplication Rule of Probabilities

The *multiplication rule* of probabilities is used to calculate the joint probability of 2 events. In other words, we are calculating the probability of these events occurring at the same time. Chapter 6 referred to this as the intersection of 2 events. For 2 independent events, the multiplication rule states the following:

$$P[A \text{ and } B] = P[A] \cdot P[B]$$

Recall from Chapter 6 that P[A and B] is also known as the joint probability of Events A and B.

For example, we can use the multiplication rule to calculate the joint probability of rolling "snake eyes" with a pair of dice. We define the events as follows:

◆ Event A: Roll a 1 on the first die.

◆ Event B: Roll a 1 on the second die.

Because these events are clearly independent, we can calculate the probability they will occur simultaneously:

$$P[A \text{ and } B] = \frac{1}{6} \cdot \frac{1}{6} = \frac{1}{36}$$

If the 2 events are dependent, thing start to heat up and the multiplication rule becomes:

$$P[A \text{ and } B] = P[A / B] \cdot P[B]$$

To demonstrate the multiplication rule with dependent events, we can go back to the tennis court and calculate P[A and B] the probability that Deb will win and that the warm-up is less than 10 minutes (from my earlier results):

$$P[B] = 0.65 \text{ and } P[A / B] = 0.31$$

$$P[A \text{ and } B] = (0.65)(0.31)$$

$$P[A \text{ and } B] = 0.20$$

Stat Facts

For dependent events, the **multiplication rule** states that $P[A \text{ and } B] = P[A / B] \cdot P[B]$. If the events are independent, the multiplication rule simplifies to $P[A \text{ and } B] = P[A] \cdot P[B]$.

Bob's Basics

The multiplication rule can be rearranged algebraically and used to calculate the conditional probability of Event B, given that Event A has occurred, with the following equation:

$$P[A/B] = \frac{P[A \text{ and } B]}{P[B]}$$

We can confirm this result by going back to the original contingency table, where we see that out of 20 matches, Deb won 4 times with a warm-up of less than 10 minutes. Therefore:

$$P[A \text{ and } B] = \frac{4}{20} = 0.20$$

Maybe Deb has a valid complaint after all. I wonder whether she ever gets tired of being right!

Mutually Exclusive Events

Two events are considered to be *mutually exclusive* if they cannot occur at the same time during the experiment. For example, suppose my experiment is to roll a single die and my events of interest are as follows:

- ◆ Event A: Roll a 1.

- ◆ Event B: Roll a 2.

Because there is no way for both of these events to occur simultaneously, they are considered to be mutually exclusive. The following Venn diagram shows Events A and B as mutually exclusive (Figure 7.1).

Figure 7.1

Events A and B are mutually exclusive.

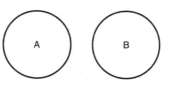

Stat Facts

Two events are considered to be **mutually exclusive** if they cannot occur at the same time during the experiment.

Events that can occur at the same time are, you guessed it, not mutually exclusive. In our tennis example, Events A and B are not mutually exclusive because Deb can win the match (A) and the warm-up can be less than 10 minutes (B) in the same experiment. The following Venn diagram shows Events A and B, which are not mutually exclusive (Figure 7.2).

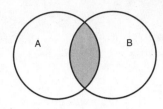

Figure 7.2

Events A and B are not mutually exclusive.

The overlapping area between the 2 circles indicates that both Events A and B can occur at the same time.

Addition Rule of Probabilities

The *addition rule* of probabilities is used to calculate the probability of the union of events—that is, the probability that either Event A or Event B will occur. For 2 events that are mutually exclusive, the addition rule states the following:

P[A or B] = P[A] + P[B]

As an example, for the single-die experiment with mutually exclusive events:

◆ Event A: Roll a 1.

◆ Event B: Roll a 2.

Stat Facts

For mutually exclusive events, the **addition rule** states that P[A or B] = P[A] + P[B]. If the events are not mutually exclusive, the addition rule becomes P[A or B] = P[A] + P[B] – P[A and B].

The simple probabilities are as follows:

$$P[A] = \frac{1}{6} = 0.167 \text{ and } P[B] = \frac{1}{6} = 0.167$$

The probability that either a 1 or a 2 will be rolled is as follows:

P[A or B] = P[A] + P[B]

P[A or B] = 0.167 + 0.167

P[A or B] = 0.334

For events that are not mutually exclusive, the addition rule states the following:

P[A or B] = P[A] + P[B] – P[A and B]

Going back to the tennis court, where …

◆ Event A = Deb wins the tennis match.

◆ Event B = The warm-up time is less than 10 minutes.

Recall that:

P[A] = 0.45 and P[B] = 0.65

P[A and B] = 0.20

Therefore, the probability that Deb will either win the match or the warm-up will be less than 10 minutes is as follows:

P[A or B] = P[A] + P[B] – P[A and B]

P[A or B] = 0.45 + 0.65 – 0.20

P[A or B] = 0.90

The logic behind subtracting P[A and B] in the addition rule is to avoid double counting. This can be demonstrated in the following table, which converts the frequency distribution to a relative frequency distribution.

Relative Frequency Distribution for Tennis Matches

Warm-Up Time	Deb Wins	Bob Wins	Total
Less than 10	4/20 = 0.20	9/20 = 0.45	13/20 = 0.65
10 or more	5/20 = 0.25	2/20 = 0.10	7/20 = 0.35
Total	9/20 = 0.45	11/20 = 0.55	20/20 = 1.00

The union of Events A and B can be displayed using Figure 7.3.

Figure 7.3

The union of Events A and B.

Warm-up Time	Deb Wins	Bob Wins	Totals
Less than 10 min	0.20	0.45	0.65
10 min or more	0.25	0.10	0.35
Totals	0.45	0.55	1.00

Bob's Basics

When converting frequencies to relative frequencies in a contingency table, always divide each number in the table by the total number of observations. In the previous example, that is 20 matches.

The probability of Deb winning the match (Event A) is represented by the box in the first column. The probability of having a warm-up of less than 10 minutes (Event B) is represented by the box in the first row. If we add P[A] + P[B], which would be the column plus the row in Figure 7.3, we are double counting P[A and B] = 0.20 and therefore need to subtract this in the addition rule for events that are not mutually exclusive.

Summarizing Our Findings

Before moving on to the last section on probability for this chapter, let's step back and take a look at what we've done so far. Figure 7.4 shows the simple, joint, and conditional probabilities in the relative frequency distribution for our tennis matches.

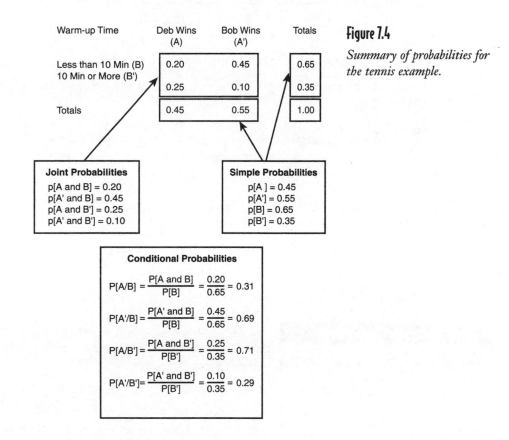

Figure 7.4

Summary of probabilities for the tennis example.

Note that:

- ◆ Event A' = Bob wins the match.

- ◆ Event B' = The warm-up is 10 minutes or more.

These conditional probabilities have revealed my secret to success on the court. The probability of me winning after a short warm-up, P[A' / B], is 0.69; whereas the probability of my opponent winning after a longer warm-up, P[A' / B'], is 0.29. I knew I should have picked another example for this chapter.

Bayes' Theorem

Thomas Bayes (1701–1761) developed a mathematical rule that deals with calculating P[B / A] from information about P[A / B]. *Bayes' theorem* states the following:

$$P[B / A] = \frac{P[B] \cdot P[A / B]}{(P[B] \cdot P[A / B] + P[B] \cdot P[A' / B'])}$$

where:

P[B'] = the probability of the complement of Event B

P[A / B'] = the probability of Event A, given that the complement to Event B has occurred

Now that looks like a mouthful, but applying it in our tennis example will clear things up. With Bayes' theorem, we can calculate P[B / A], which is the probability that the warm-up was less than 10 minutes given that Deb won the match. Using the values from the previous figure:

$$P[B/A] = \frac{0.65 \cdot 0.31}{(0.65 \cdot 0.31) + (0.35 \cdot 0.71)}$$

$$P[B/A] = \frac{0.20}{0.20 + 0.25}$$

$$P[B/A] = 0.44$$

Random Thoughts
Not only was Thomas Bayes a prominent mathematician, he was also a published Presbyterian minister who used mathematics to study religion.

Knowing that Deb won the match, we can say there is a 44 percent chance that the warm-up was less than 10 minutes.

We can confirm this result by looking at the original contingency table. Because Deb won 9 matches and from those, 4 had a warm-up of less than 10 minutes:

$$P[B/A] = \frac{4}{9} = 0.44$$

Ta da! Please hold your applause until the end of the book.

Your Turn

A political telephone survey of 260 people asked whether they were in favor or not in favor of a proposed law. Each person was identified as Republican or Democrat. The results are shown in the following contingency table.

Party	In Favor	Not in Favor	Total
Republican	98	54	152
Democrat	79	29	108
Total	177	83	260

A person from the survey is selected at random. We define:

♦ Event A: The person selected is in favor of the new law.

♦ Event B: The person selected is a Republican.

1. Determine the probability that the selected person is in favor of the new law.

2. Determine the probability that the selected person is a Republican.

3. Determine the probability that the selected person is not in favor of the new law.

4. Determine the probability that the selected person is a Democrat.

5. Determine the probability that the selected person is in favor of the new law given that the person is a Republican.

6. Determine the probability that the selected person is not in favor of the new law given that the person is a Republican.

7. Determine the probability that the selected person is in favor of the new law given that the person is a Democrat.

8. Determine the probability that the selected person is in favor of the new law and that the person is a Republican.

9. Determine the probability that the selected person is in favor of the new law and that the person is a Democrat.

10. Determine the probability that the selected person is in favor of the new law or that the person is a Republican.

11. Determine the probability that the selected person is in favor of the new law or that the person is a Democrat.

12. Using Bayes' theorem, calculate the probability that the selected person was a Republican, given that the person was in favor of the new law.

The Least You Need to Know

♦ Conditional probability is defined as the probability of Event A knowing that Event B has already occurred.

♦ Events A and B are said to be independent of each other if the occurrence of Event B has no effect on the probability of Event A. If Events A and B are not independent of one another, then they are said to be dependent events.

♦ For dependent events, the multiplication rule states that $P[A \text{ and } B] = P[A/B] \cdot P[B]$. If the events are independent, the multiplication rule simplifies to $P[A \text{ and } B] = P[A] \cdot P[B]$.

♦ Two events are considered to be mutually exclusive if they cannot occur at the same time during the experiment.

♦ For mutually exclusive events, the addition rule states that $P[A \text{ or } B] = P[A] + P[B]$. If the events are not mutually exclusive, the addition rule becomes $P[A \text{ or } B] = P[A] + P[B] - P[A \text{ and } B]$.

♦ Bayes' theorem deals with calculating $P[B / A]$ from information about $P[A / B]$ using the following formula: $P[B/A] = \dfrac{P[B] \cdot P[A/B]}{(P[B] \cdot P[A/B] + P[B'] \cdot P[A/B'])}$

Chapter 8

Counting Principles and Probability Distributions

In This Chapter

◆ Using the fundamental counting principle

◆ Distinguishing between permutations and combinations

◆ Defining a random variable and probability distribution

◆ Calculating the mean and variance of a discrete probability distribution

Well, we've finally arrived at our third and last chapter on general probability concepts. This chapter will set the stage for the last 3 chapters in Part 2, which focus on specific types of probability distributions. Before you know it, we'll be knee deep with inferential statistics.

This chapter will also teach you how to count. This type of counting, however, goes far beyond what you've seen on *Sesame Street*. Counting events is an important step in calculating probabilities and must be done with care.

Counting Principles

To use classical probability, which was introduced way back in Chapter 6, we need to be able to count the number of events of interest along with the total number of events that are possible in the sample space. For simple events, like rolling a single die, the number of possible outcomes (6) is obvious. But for more complex events, like a state lottery drawing, we need to rely on techniques known as counting principles to arrive at the correct answer. The next 3 sections discuss these techniques.

The Fundamental Counting Principle

After a tough round of golf on a hot afternoon, Brian, John, and I decide to revive our spirits at the ice cream store on the way home. There I'm overwhelmed with deciding between 4 flavors and 3 toppings to indulge in. How many different combinations of ice cream and toppings am I faced with? The *fundamental counting principle* comes to my rescue by telling me that if 1 event (my ice cream choice) can occur in m ways and a second event (my topping choice) can occur in n ways, the total number of ways both events can occur together is $m*n$ ways. In my case, I have $4*3=12$ combinations of flavors and toppings in which to blow my diet. (I'll leave that topic for another chapter.)

This principle can be extended to more than two events. In addition to flavors and toppings, I have another tempting choice between a small and large serving. That leaves me with the mind-boggling decision with $4*3*2=24$ combinations, which are summarized in the table that follows my list of options.

Stat Facts

According to the **fundamental counting principle**, if 1 event can occur in m ways and a second event can occur in n ways, the total number of ways both events can occur together is $m*n$ ways. This principle can be extended to more than two events.

Ice Cream Flavors

CH = Death by Chocolate

VA = Vanilla

ST = Strawberry

CF = Coffee

Toppings

HF = Hot Fudge

BS = Butterscotch

SP = Sprinkles

Sizes

SM = Small

LG = Large

List of Combinations (Flavor-Topping-Size)			
CH-HF-LG	VA-HF-LG	ST-HF-LG	CF-HF-LG
CH-HF-SM	VA-HF-SM	ST-HF-SM	CF-HF-SM
CH-BS-LG	VA-BS-LG	ST-BS-LG	CF-BS-LG
CH-BS-SM	VA-BS-SM	ST-BS-SM	CF-BS-SM
CH-SP-LG	VA-SP-LG	ST-SP-LG	CF-SP-LG
CH-SP-SM	VA-SP-SM	ST-SP-SM	CF-SP-SM

Can you guess which choice a certain chocolate-loving author made?

Another demonstration of the fundamental counting principle is to calculate the number of unique combinations for a state's automobile license plates. Suppose the state plates have 3 letters followed by 4 numbers. The number 0 and the letter O are not eligible because their resemblance may cause confusion. Because there are 25 possible letters and 9 possible numbers, the total number of unique combinations is as follows:

First Letter	Second Letter	Third Letter	First Number	Second Number	Third Number	Fourth Number
25*	25*	25*	9*	9*	9*	9 =

That's 102,515,625 possible combinations!

Permutations

Permutations are the number of different ways in which objects can be arranged in order. In a permutation, each item appears only once. The number of permutations of n distinct objects is $n!$ (expressed as n factorial) and is defined as follows:

$$n! = n * (n-1) * (n-2) * (n-3) * \cdots * 4 * 3 * 2 * 1$$

Stat Facts

Permutations are the number of different ways in which objects can be arranged in order. The number of permutations of n objects taken r at a time can be found by

$$_n P_r = \frac{n!}{(n-r)!}.$$

By definition, $0! = 1$. For instance, $6! = 6*5*4*3*2*1 = 720$. As an example, there are 6 permutations for the numbers 1, 2, and 3, as shown here:

123 132 213 231 312 321

Because:

$3! = 3*2*1 = 6$

Before the beginning of a professional basketball game, the starting 5 players are announced 1 at a time. How many different ways can the order that the players are announced be arranged? The number of permutations is $5! = 5*4*3*2*1 = 120$.

Suppose we want to select only some of the objects in the group. The number of permutations of n objects taken r at a time can be found as follows:

$$_n P_r = \frac{n!}{(n-r)!}$$

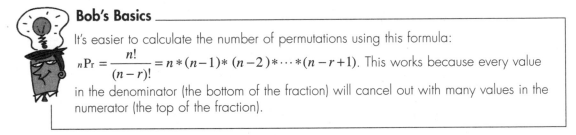

Bob's Basics

It's easier to calculate the number of permutations using this formula:

$_n P_r = \dfrac{n!}{(n-r)!} = n*(n-1)*(n-2)*\cdots*(n-r+1)$. This works because every value in the denominator (the bottom of the fraction) will cancel out with many values in the numerator (the top of the fraction).

Going back to our basketball example, if there are 12 players on the team, how many different ways can any 5 players on the team be announced to start the game? In this case, because $n = 12$ and $r = 5$, the number of permutations is as follows:

$$_{12} P_5 = \frac{12!}{(12-5)!} = \frac{12*11*10*9*8*7*6*5*4*3*2*1}{7*6*5*4*3*2*1}$$

$$_{12} P_5 = \frac{12!}{(12-5)!} = 12*11*10*9*8 = 95,040$$

I'm sure glad it's not my job to decide who gets announced first.

Sometimes the order of events is not of consequence. Those cases are discussed in the next section.

Combinations

Combinations are similar to permutations, except that the order of the objects is not important. The number of combinations of n objects taken r at a time can be found as follows:

$$_nC_r = \frac{n!}{(n-r)\,r!}$$

Stat Facts

Combinations are the number of different ways in which objects can be arranged without regard to order. The number of combinations of n objects taken r at a time can be found by $_nC_r = \dfrac{n!}{(n-r)\,r!}$.

Bob's Basics

It's easier to calculate the number of combinations using this formula:

$$_nC_r = \frac{n!}{(n-r)!\,r!} = \frac{n*(n-1)*(n-2)*\cdots*(n-r+1)}{r!}$$ using the same logic as the permutation formula.

For example, in poker, 5 cards are selected randomly from a deck of 52 cards. How many 5-card combinations exist?

$$_{52}C_5 = \frac{52!}{(52-5)!\,5!} = \frac{52*51*50*49*48}{5*4*3*2*1} = 2,598,960$$

How many 5-card permutations exist?

$$_{52}P_5 = \frac{52!}{(52-5)!} = 52*51*50*49*48 = 311,875,200$$

There are more 5-card permutations because the following 2 poker hands would be considered 2 different permutations but be counted as only 1 combination because they are the same cards only in different order:

Hand 1	Hand 2
Ace of Spades	Ace of Spades
Queen of Hearts	Ten of Spades
Ten of Spades	Queen of Hearts
Ten of Diamonds	Ten of Diamonds
Three of Clubs	Three of Clubs

Now that we know the total number of 5-card combinations from a 52-card deck, we can calculate the probability of a flush, which is any 5 cards that are all the same suit (spades, clubs, hearts, or diamonds). First we need to count the number of 5-card flushes of 1 suit, let's say diamonds. Because there are 13 diamonds in the deck, the number of combinations of these 13 diamonds, taken 5 at a time is as follows:

$$_{13}C_5 = \frac{13!}{(13-5)!5!} = \frac{13*12*11*10*9}{5*4*3*2*1} = 1,287$$

Because there are 4 suits in the deck, the total number of 5-card flushes from any suit is $1287*4 = 5,148$. Therefore, the probability of being dealt a flush in a 5-card hand is:

$$P[\text{Flush}] = \frac{5,148}{2,598,960} = 0.002$$

Bob's Basics

An alternate notation for $_nC_r$ is $\binom{n}{r}$, which you may come across in other textbooks. Statisticians just love to have different notations for the same concept!

or roughly twice in 1,000 hands of poker. Ready to deal?

Combinations are also useful for calculating the probability of winning a state lottery drawing. A typical lottery game requires you to pick 6 numbers out of a possible 49. Because the order of the numbers does not matter, we use the combination rather than the permutation formula. The number of 6-number combinations from a pool of 49 numbers is this:

$$_{49}C_6 = \frac{49!}{(49-6)!6!} = \frac{49*48*47*46*45*44}{6*5*4*3*2*1} = 13,983,816$$

Because there are nearly 14 million different 6-number combinations, the probability that your combination is the winner is as follows:

$$P[\text{Winning a 6 / 49 Lottery}] = \frac{1}{13,983,816} = 0.00000007$$

Wrong Number

Probability does not have a memory. The same 6 numbers that were selected in last week's lottery drawing have the exact same probability of being chosen again in this week's lottery. That's because the two drawings are independent events and have absolutely no influence on each other. Therefore, choosing a lottery number because it has not been selected recently does not increase your odds of winning. Sorry if I ruined your favorite strategy!

With those chances of winning the lottery, you better not quit your day job just yet.

Using Excel to Calculate Permutations and Combinations

Here's something that's pretty cool—rather than deal with all those nasty factorial calculations, we can let Excel figure out the number of permutations or combinations for us. We saw earlier that $_{12}P_5 = 95,040$. We can confirm this using Excel's built-in PERMUT function, which has the following characteristics:

PERMUT(n, r)

Figure 8.1 shows the PERMUT function in action for $_{12}P_5$.

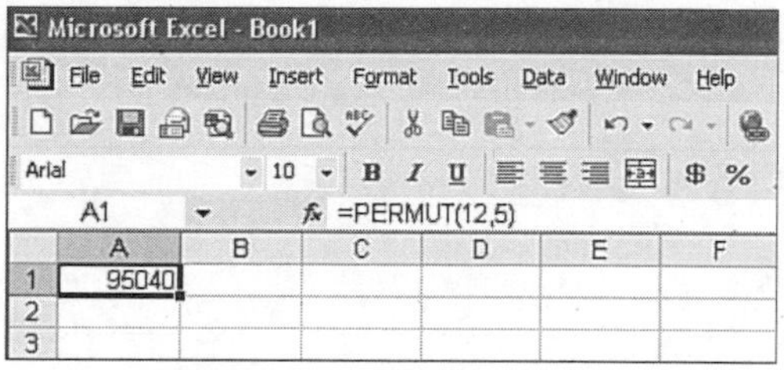

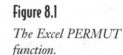

Figure 8.1

The Excel PERMUT function.

Cell A1 contains the Excel formula =PERMUT(12,5) with the result being 95,040.

We also can rely on Excel to provide the number of combinations. Recall that $_{49}C_6 = 13,983,816$. We can show this using Excel's built-in COMBIN function, which has the following characteristics:

COMBIN(n, r)

Figure 8.2 shows the COMBIN function for $_{49}C_6$.

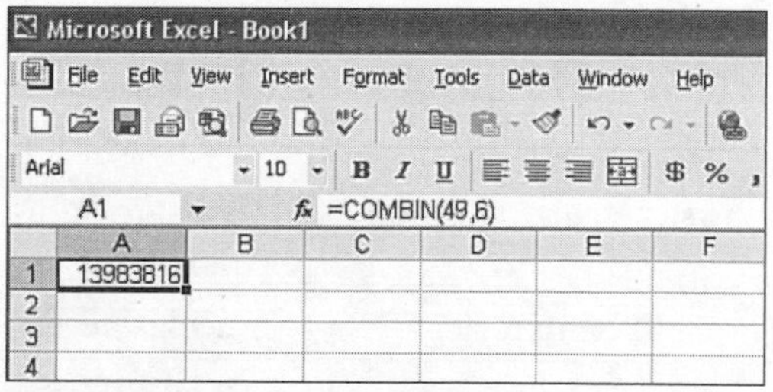

Figure 8.2

The Excel COMBIN function.

Cell A1 contains the Excel formula =COMBIN(49,6) with the result being 13983816. See! I told you it was pretty cool.

This wraps up the topic of counting principles. Many of you may have been surprised how complicated it can be to count events. But this is an important concept in statistics that we will revisit again in Chapter 9.

Probability Distributions

This section of Chapter 8 will prepare you for the last 3 chapters of Part 2. We need to lay some groundwork here about probability distributions in general before discussing specific distributions in Chapters 9, 10, and 11.

In general, a probability distribution is a listing of all the possible outcomes of an experiment along with the relative frequency or probability of each outcome. As an example, consider flipping a coin twice and recording the number of heads (H). The sample space for this experiment is shown in the following table (tails = T).

First Coin	Second Coin	Number of Heads
H	H	2
H	T	1
T	H	1
T	T	0

The following table shows the relative frequency distribution for the number of heads.

Number of Heads	Frequency	Relative Frequency
0	1	1/4 = 0.25
1	2	2/4 = 0.50
2	1	1/4 = 0.25
	Total = 4	Total = 1.00

The previous table represents the probability distribution for the number of heads after flipping a coin twice. For instance, the probability of 2 heads with 2 coin flips is 25 percent.

Before jumping into more details of probability distributions, we have some key terms to define.

Bob's Basics

Probability distributions play a major role in the use of inferential statistics. A firm understanding of this topic is vital to the success of mastering statistics.

Random Variables

In Chapter 6, we talked about conducting experiments to acquire data. Examples of experiments could be rolling dice or playing a tennis match with your spouse. The outcomes of these experiments are considered *random variables*. By definition, these outcomes are not known with certainty before the experiment. But with the help of probability theory and statistics, we can often make a statement about the probability of certain outcomes. For example, I can't predict with certainty the result of rolling a single die, but I do know that the probability of rolling a 1 is 1/6. In this case, the random variable would be the number that I would roll.

Stat Facts

A **random variable** is an outcome that takes on a numerical value as a result of an experiment. The value is not known with certainty before the experiment. The value of the random variable is often denoted by x. For instance, with the previous single die example, we would say $P[x = 1] = \dfrac{1}{6}$.

All random variables are not created equal. The first type is known as *continuous* random variables, which are the result of a measurement on a continuous number scale. For example, each morning when I take a deep breath and step on the bathroom scale to weigh myself (taking a deep breath and holding it somehow makes me feel lighter), I'm looking down in shock and disbelief at a continuous random variable. (Maybe I should have chosen the small Death by Chocolate serving.) Examples of values for continuous random variables of this sort could be 180, 181.5, 183.2, and so on. (I'll stop there.) Because this is a continuous variable, my morning weight could take on an unlimited number of possible values, which is a very disconcerting thought.

The second type of random variable is *discrete*; discrete random variables are the result of counting outcomes rather than measuring them. Discrete random variables can only take on a certain number of integer values within an interval. An example of a discrete random variable would be my golf score for my next round, because this

Stat Facts

A random variable is **continuous** if it can assume any numerical value within an interval as a result of measuring the outcome of an experiment. A random variable is **discrete** if it is limited to assuming only specific integer values as a result of counting the outcome of an experiment.

value is arrived at by counting my total strokes over 18 holes of play. Obviously, this value needs to be an integer, such as 94, because there is no way to count a partial stroke (even though there are times my golf swing feels like one).

Additional examples of continuous random variables include the following:

♦ The amount of local rainfall, in inches, this month

♦ The length of time a customer required at a checkout lane in the grocery store

♦ The speed of a vehicle traveling on the interstate measured by a radar gun

Additional examples of discrete variables include the following:

♦ The number of days during the month in which it rained

♦ The number of customers standing in line waiting to check out at the grocery store

♦ The number of cars that were found driving faster than the speed limit during the past hour

Continuous random variables will be discussed in more detail in Chapter 11. The remainder of this chapter and Chapters 9 and 10 will focus solely on discrete random variables.

Discrete Probability Distributions

A listing of all the possible outcomes of an experiment for a discrete random variable along with the relative frequency or probability of each outcome is called a discrete probability distribution. To illustrate this concept, I'll use the following example.

My oldest daughter, Christin, was a very accomplished competitive swimmer between the ages of 7 and 13. Her talent certainly didn't come from my side of the family. The following table is a relative frequency distribution showing the number of first-, second-, third-, fourth-, and fifth-place finishes over 50 races.

Place	Number of Races	Relative Frequency (Probability)
1	27	27/50 = 0.54
2	12	12/50 = 0.24
3	7	7/50 = 0.14
4	3	3/50 = 0.06
5	1	1/50 = 0.02
	Total = 50	Total = 1.00

If we define the random variable x = the place Christin finished in a race, the previous table would be the discrete probability distribution for the variable x. From this table, we can state the probability that Christin will finish first as follows:

$P[x = 1] = 0.54$

Or we can state the probability that Christin will finish either first or second as follows:

$P[x = 1 \text{ or } x = 2] = 0.54 + 0.24 = 0.78$

Figure 8.3 shows the discrete probability distribution for x graphically.

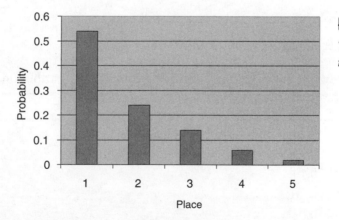

Figure 8.3

The discrete probability distribution for Christin's races.

Rules for Discrete Probability Distributions

Any discrete probability distribution needs to meet the following requirements:

◆ Each outcome in the distribution needs to be mutually exclusive—that is, the value of the random variable cannot fall into more than 1 of the frequency distribution classes. For example, it is not possible for Christin to take first and second place in the same race.

♦ The probability of each outcome, $P[x]$, must be between 0 and 1; that is, $0 \leq P[x] \leq 1$ for all values of x. In the previous example, $P[x = 3] = 0.14$, which falls between 0 and 1.

♦ The sum of the probabilities for all the outcomes in the distribution needs to add up to 1; that is, $\sum_{i=1}^{n} P[x_i] = 1$. In the swimming example:

$P[x = 1] + P[x = 2] + P[x = 3] + P[x = 4] + P[x = 5] = 0.54 + 0.24 + 0.14 + 0.06 + 0.02 = 1.00$

The Mean of a Discrete Probability Distribution

The mean of a discrete probability distribution is simply a weighted average (discussed in Chapter 4) calculated using the following formula:

$$\mu = \sum_{i=1}^{n} x_i * P[x_i]$$

where:

μ = the mean of the discrete probability distribution

x_i = the value of the random variable for the ith outcome

$P[x_i]$ = the probability of the ith outcome will occur

n = the number of outcomes in the distribution

The table that follows revisits Christin's swimming probability distribution.

Place x_i	Probability $P[x_i]$
1	0.54
2	0.24
3	0.14
4	0.06
5	0.02

The mean of this discrete probability distribution is as follows:

$$\mu = \sum_{i=1}^{n} x_i * P[x_i] = (1 \cdot 0.54) + (2 \cdot 0.24) + (3 \cdot 0.14) + (4 \cdot 0.06) + (5 \cdot 0.02)$$

$\mu = 1.78$

This mean is telling us that Christin's average finish for a race is 1.78 place! How does she do that? Obviously, this will never be the result of any one particular race. Rather, it represents the average finish of many races. The mean of a discrete probability distribution does not have to equal 1 of the values of the random variable (1, 2, 3, 4, or 5 in this case).

Another term for describing the mean of a probability distribution is the *expected value*, E[x]. Therefore:

$$E[x] = \mu = \sum_{i=1}^{n} x_i * P[x_i]$$

Didn't I say statisticians love all sorts of notation to describe the same concept!

Stat Facts

An **expected value** is the mean of a probability distribution.

The Variance and Standard Deviation of a Discrete Probability Distribution

Just when you thought it was safe to get back into the water, along comes another variance! Well, if you've seen 1 variance calculation, you've seen them all. The variance for a discrete probability distribution can be calculated as follows:

$$\sigma^2 = \sum_{i=1}^{n} (x_i - \mu)^2 * P[x_i]$$

where:

σ^2 = the variance of the discrete probability distribution

As before, the standard deviation of the distribution is as follows:

$$\sigma = \sqrt{\sigma^2}$$

To demonstrate the use of these equations, we'll rely on Christin's swimming distribution. The calculations are summarized in the following table.

x_i	$P[x_i]$	μ	$x_i - \mu$	$(x_i - \mu)^2$	$(x_i - \mu)^2 \cdot P[x_i]$
1	0.54	1.78	−0.78	0.608	0.328
2	0.24	1.78	0.22	0.048	0.012
3	0.14	1.78	1.22	1.488	0.208
4	0.06	1.78	2.22	4.928	0.296
5	0.02	1.78	3.22	10.368	0.208

$$\sigma^2 = \sum_{i=1}^{n} (x_i - \mu)^2 * P[x_i] = 1.052$$

The standard deviation of this distribution is this:

$$\sigma = \sqrt{\sigma^2} = \sqrt{1.052} = 1.026$$

A more efficient way to calculate the variance of a discrete probability distribution is as follows:

$$\sigma^2 = \left(\sum_{i=1}^{n} x_i^2 * P[x_i] \right) - \mu^2$$

The following table summarizes these calculations using Christin's swimming example.

x_i	$P[x_i]$	x_i^2	$x_i^2 * P[x_i]$
1	0.54	1	0.54
2	0.24	4	0.96
3	0.14	9	1.26
4	0.06	16	0.96
5	0.02	25	0.50

$$\sum_{i=1}^{n} x_i^2 * P[x_i] = 4.22$$

$$\sigma^2 = \left(\sum_{i=1}^{n} x_i^2 * P[x_i] \right) - \mu^2$$

$$\sigma^2 = 4.22 - (1.78)^2$$

$$\sigma^2 = 1.052$$

As you can see, the result is the same, but with less effort!

Your Turn

1. A restaurant has a menu with 3 appetizers, 8 entrées, 4 desserts, and 3 drinks. How many different meals can be ordered?

2. A multiple-choice test has 10 questions with each question having 4 choices. What is the probability that a student, who randomly answers each question, will answer each question correctly?

3. The NBA teams with the 13 worst records at the end of the season participate in a lottery to determine the order in which they will draft new players for the next season. How many different arrangements exist for the drafting order for these 13 teams?

4. In a race with 8 swimmers, how many ways can swimmers finish first, second, and third?

5. How many different ways can 10 new movies be ranked first and second by a movie critic?

6. A combination lock has a total of 40 numbers and will unlock with the proper 3-number sequence. How many possible combinations exist?

7. I would like to select 3 paperback books from a list of 12 books to take on vacation. How many different sets of 3 books can I choose?

8. A panel of 12 jurors needs to be selected from a group of 50 people. How many different juries can be selected?

9. A survey of 450 families was conducted to find how many cats were owned by each respondent. The following table summarizes the results.

Number of Cats	Number of Families
0	137
1	160
2	112
3	31
4	10

Develop a probability distribution for this data and calculate the mean, variance, and standard deviation.

The Least You Need to Know

♦ The fundamental counting principle states that if 1 event can occur in m ways and a second event can occur in n ways, the total number of ways both events can occur together is $m * n$ ways. This principle can be extended to more than 2 events.

♦ Permutations are the number of different ways in which objects can be arranged in order. Combinations are the number of different ways in which objects can be arranged when order is of no importance.

♦ A probability distribution is a listing of all the possible outcomes of an experiment along with the relative frequency or probability of each outcome.

- A random variable is an outcome that takes on a numerical value as a result of an experiment. The value is not known with certainty before the experiment.

- A random variable is continuous if it can assume any numerical value within an interval as a result of measuring the outcome of an experiment. A random variable is discrete if it is limited to assuming only specific integer values as a result of counting the outcome of an experiment.

- The mean of a discrete probability distribution is found as follows:

$$\mu = \sum_{i=1}^{n} x_i * P[x_i] .$$

- The variance of a discrete probability distribution is found as follows:

$$\sigma^2 = \sum_{i=1}^{n} (x_i - \mu)^2 * P[x_i] .$$

9

The Binomial Probability Distribution

In This Chapter

- ◆ Describe the characteristics of a binomial experiment
- ◆ Calculate the probabilities for a binomial distribution
- ◆ Find probabilities using a binomial table
- ◆ Find binomial probabilities using Excel
- ◆ Calculate the mean and standard deviation of a binomial distribution

Our discussion of discrete probability distributions so far has been limited to general distributions based on historical data that has been previously collected. There are, however, theoretical probability distributions that are based on a mathematical formula rather than historical data. The first of these, the binomial probability distribution, will be addressed in this chapter.

There are many types of problems where we are interested in the probability of an event occurring several times. A classical example that has been torturing students for many years is "What is the chance of getting 7 heads when tossing a coin 10 times?" By the time you are finished with this chapter, answering this question will be a piece of cake!

Characteristics of a Binomial Experiment

If your memory serves you correctly, we defined experiments in Chapter 6 as the process of measuring or observing an activity for the purpose of collecting data. Let's say our experiment of interest involves a certain professional basketball player shooting free throws. Each free throw would be considered a *trial* for the experiment. For this particular experiment, there are only 2 possible outcomes for each trial; either the free throw goes in the basket (a success) or it doesn't (a failure). Because there are only 2 possible outcomes for each trial, this is known as a *binomial experiment*.

Stat Facts

A **binomial experiment** has the following characteristics: (1) The experiment consists of a fixed number of trials denoted by *n*; (2) each trial has only 2 possible outcomes, a success or a failure; (3) the probability of success and the probability of failure are constant throughout the experiment; (4) each trial is independent of any other trial in the experiment.

Random Thoughts

Binomial experiments are also known as a Bernoulli process, named after Swiss mathematician James Bernoulli, who lived during the 1600s. Repeating a Bernoulli process several times is referred to as Bernoulli trials. This concept has been haunting students for hundreds of years!

Let's say that our player of interest is Michael Jordan, who, historically, has made 80 percent of his free throws. So the probability of success, *p*, of any given free throw is 0.80. Because there are only 2 outcomes possible, the probability of failure for any given free throw, *q*, is 0.20. For a binomial experiment, the values of *p* and *q* must be the same for every trial in the experiment. Because only 2 outcomes are allowed in a binomial experiment, $p = 1 - q$ always holds true.

Finally, a binomial experiment requires that each trial is independent of any other trials. In other words, the probability of the second free throw being successful is not affected by whether the first free throw was successful. Other examples of binomial experiments include the following:

- Testing whether a part is defective after it has been manufactured

- Observing the number of correct responses in a multiple-choice exam

- Counting the number of American households that have an Internet connection

Now that we have defined the ground rules for binomial experiments, we are ready to graduate to calculating binomial probabilities.

The Binomial Probability Distribution

The binomial probability distribution allows us to calculate the probability of a specific number of successes for a certain number of trials. Therefore, the random variable for this distribution would be the number of successes that were observed. To demonstrate a binomial distribution, I will use the following example.

Deb had trained our dog, Kaylee, to do an incredible trick. First thing every morning after she lets the dog out the back door, Kaylee runs like greased lightning around the house, down our rather long driveway, grabs our newspaper, races to the back door where she dutifully deposits it on the step. In return for this vital chore for our household, she gets 2 cups of dry dog food in a plastic bowl. Amazing, you say. But you've only heard the half of it. Somehow, in the tiny recesses of Kaylee's doggy brain, she has worked out the remarkable deduction that "2 breakfasts are better than 1" and at every opportunity goes on a neighborhood hunt for more newspapers to deposit on our back step. Once she dragged an entire *phone book* back, thinking maybe this would earn her a bonus. We have failed miserably trying to train Kaylee to return these papers—apparently tiny doggy brains don't work in reverse.

So my job on many afternoons is to carefully return the stolen merchandise, hoping my neighbors fail to notice the dog slobber on their 3-day-old paper. Anyway, let's say on any particular day there is a 30 percent probability that Kaylee will bring back 1 stolen paper and a 70 percent chance that she won't. We will assume that she will not bring back more than 1 paper a day. This scenario represents a binominal experiment with each day being a Bernoulli trial with $p = 0.30$ (the probability of a "success") and $q = 0.70$ (the probability of a "failure"). We can calculate the probability of r successes in n trials using the binomial distribution, as follows:

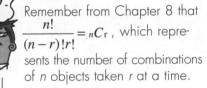

Bob's Basics

Remember from Chapter 8 that $\frac{n!}{(n-r)!r!} = {}_nC_r$, which represents the number of combinations of n objects taken r at a time.

$$P[r,n] = \frac{n!}{(n-r)!r!}p^r q^{n-r}$$

With this equation, we can calculate the probability that Kaylee will bring back 3 papers over the next 5 days.

$$P[3,5] = \frac{5!}{(5-3)!3!}(0.3)^3(0.7)^{5-3}$$

$$P[3,5] = \left(\frac{120}{2 \cdot 6}\right)(0.027)(0.49) = 0.1323$$

There is a 13 percent chance that the neighborhood paper bandit will strike 3 times during the next 5 days. We can also calculate the probability that she will round up 0, 1, 2, 4, or 5 papers over the next 5 days.

Bob's Basics

Remember from Chapter 8, $0! = 1$. Also $x^0 = 1$ for any value of x.

For $r = 0$:

$$P[0,5] = \frac{5!}{(5-0)!0!}(0.3)^0(0.7)^{5-0}$$

$$P[0,5] = \left(\frac{120}{120 \cdot 1}\right)(1)(0.1681) = 0.1681$$

For $r = 1$:

$$P[1,5] = \frac{5!}{(5-1)!1!}(0.3)^1(0.7)^{5-1}$$

$$P[1,5] = \left(\frac{120}{24 \cdot 1}\right)(0.3)(0.2401) = 0.3601$$

For $r = 2$:

$$P[2,5] = \frac{5!}{(5-2)!2!}(0.3)^2(0.7)^{5-2}$$

$$P[2,5] = \left(\frac{120}{6 \cdot 2}\right)(0.09)(0.343) = 0.3087$$

For $r = 4$:

$$P[4,5] = \frac{5!}{(5-4)!4!}(0.3)^4(0.7)^{5-4}$$

$$P[4,5] = \left(\frac{120}{1 \cdot 24}\right)(0.0081)(0.7) = 0.0283$$

For $r = 5$:

$$P[5,5] = \frac{5!}{(5-5)!5!}(0.3)^5(0.7)^{5-5}$$

$$P[5,5] = \left(\frac{120}{1 \cdot 120}\right)(0.0024)(1) = 0.0024$$

The following table summarizes all the previous probabilities.

r	P[r,5]
0	0.1681
1	0.3601
2	0.3087

r	P[r,5]
3	0.1323
4	0.0283
5	0.0024
	Total = 1.0

This table represents the binomial probability distribution for r successes in 5 trials with the probability of success equal to 0.30. Notice that the sum of all the probabilities equals 1, which is a requirement for all probability distributions. Figure 9.1 shows this probability distribution as a histogram.

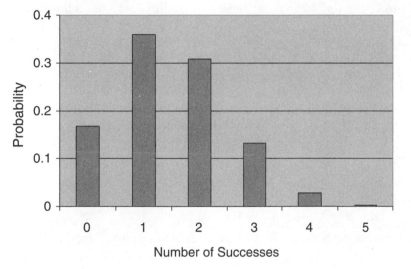

Figure 9.1

Binomial probability distribution.

From this figure, we can see the most likely number of papers that Kaylee will show up with over 5 days is 1.

Finally, we can calculate the probability of multiple events for this distribution. For instance, the probability that Kaylee will steal at least 3 papers over the next 5 days is this:

$$P[r \geq 3] = P[3,5] + P[4,5] + P[5,5]$$

$$P[r \geq 3] = 0.1323 + 0.0283 + 0.0024 = 0.163$$

Also, the probability that Kaylee will take no more than 1 paper over the next 5 days is this:

$$P[r \leq 1] = P[0,5] + P[1,5]$$

$$P[r \leq 1] = 0.1684 + 0.3601 = 0.5285$$

Our neighbors will be thrilled to see these figures!

Binomial Probability Tables

As the number of trials increases in a binomial experiment, calculating probabilities using the previous formula will really drain the batteries in your calculator and possibly even your brain. An easier way to arrive at these probabilities is to use a binomial probability table, which I have conveniently provided in Appendix B of this book. Below is an excerpt from this appendix with the probabilities from our previous example underlined.

The probability table is organized by values of n, the total number of trials. The number of successes, r, are the rows of each section, whereas the probability of success, p, are the columns. Notice that the sum of each block of probabilities for a particular value of p adds to 1.0.

Values of p

n	r	0.1	0.2	0.3	0.4	0.5	0.6	0.7	0.8	0.9
4	0	0.6561	0.4096	0.2401	0.1296	0.0625	0.0256	0.0081	0.0016	0.0001
	1	0.2916	0.4096	0.4116	0.3456	0.2500	0.1536	0.0756	0.0256	0.0036
	2	0.0486	0.1536	0.2646	0.3456	0.3750	0.3456	0.2646	0.1536	0.0486
	3	0.0036	0.0256	0.0756	0.1536	0.2500	0.3456	0.4116	0.4096	0.2916
	4	0.0001	0.0016	0.0081	0.0256	0.0625	0.1296	0.2401	0.4096	0.6561

Values of p

n	r	0.1	0.2	0.3	0.4	0.5	0.6	0.7	0.8	0.9
5	0	0.5905	0.3277	<u>0.1681</u>	0.0778	0.0313	0.0102	0.0024	0.0003	0.0000
	1	0.3280	0.4096	<u>0.3601</u>	0.2592	0.1563	0.0768	0.0284	0.0064	0.0005
	2	0.0729	0.2048	<u>0.3087</u>	0.3456	0.3125	0.2304	0.1323	0.0512	0.0081
	3	0.0081	0.0512	<u>0.1323</u>	0.2304	0.3125	0.3456	0.3087	0.2048	0.0729

n	r	0.1	0.2	0.3	0.4	0.5	0.6	0.7	0.8	0.9
4		0.0005	0.0064	0.0283	0.0768	0.1563	0.2592	0.3601	0.4096	0.3281
5		0.0000	0.0003	0.0024	0.0102	0.0313	0.0778	0.1681	0.3277	0.5905

One limitation of using binomial tables is that you are restricted to using only the values of p that are shown in the table. For instance, the previous table would not be useful for p = 0.35. Other statistics books may contain binomial tables that are more extensive than the one in Appendix B.

Using Excel to Calculate Binomial Probabilities

A convenient way to calculate binomial probabilities is to rely on our friend Excel, with its BINOMDIST function. This built-in function has the following characteristics:

BINOMDIST(r, n, p, cumulative)

where:

cumulative = FALSE if you want the probability of exactly r successes

cumulative = TRUE if you want the probability of r or fewer successes

For instance, Figure 9.2 shows the BINOMDIST function being used to calculate the probability that Kaylee will bring back exactly 2 papers during the next 5 days.

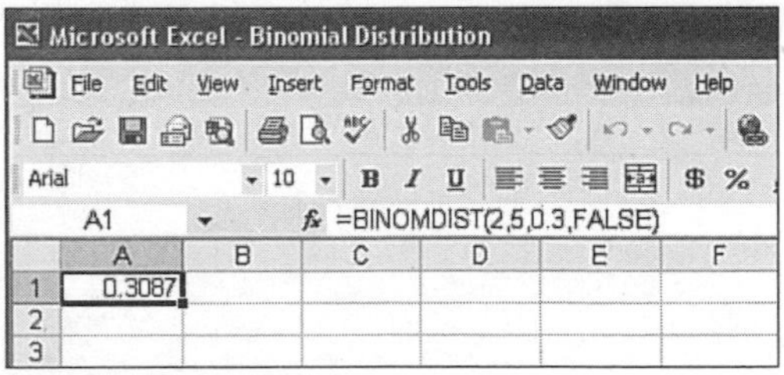

Figure 9.2

BINOMDIST function in Excel for exactly r successes.

Cell A1 contains the Excel formula =BINOMDIST(2,5,0.3,FALSE) with the result being 0.3087.

Excel will also calculate the probability that Kaylee will bring back no more than 2 papers over the next 5 days, as shown in Figure 9.3.

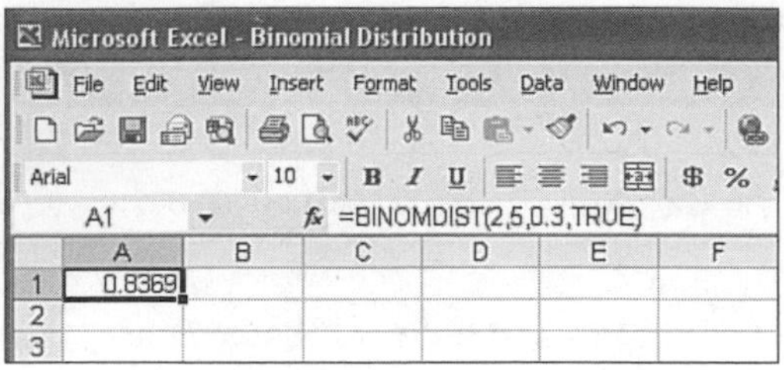

Figure 9.3

BINOMDIST function in Excel for no more than r successes.

Cell A1 contains the Excel formula =BINOMDIST(2,5,0.3,TRUE) with the result being 0.8369, which is the same as this:

$$P[r \leq 2] = P[0,5] + P[1,5] + P[2,5]$$

$$P[r \leq 2] = 0.1681 + 0.3601 + 0.3087 = 0.8369$$

In other words, there is more than an 83 percent chance Kaylee will show up at our back door with 0, 1, or 2 papers that don't belong to us during the next 5 days. That dog sure does keep me busy.

One benefit of using Excel to determine binomial probabilities is that you are not limited to the values of p shown in the binomial table in Appendix B. Excel's BINOMDIST function allows you to use any value between 0 and 1 for p.

The Mean and Standard Deviation for the Binomial Distribution

The mean for a binomial probability distribution can be calculated using the following equation:

$$\mu = np$$

where:

n = the number of trials

p = the probability of a success

For Kaylee's example, the mean of the distribution is as follows:

$$\mu = np = (5)(0.3) = 1.5 \text{ papers}$$

In other words, Kaylee brings back, on average, 1.5 papers every 5 days.

The standard deviation for a binomial probability distribution can be calculated using the following equation:

$$\sigma = \sqrt{npq}$$

where:

q = the probability of a failure

For our example, the standard deviation of the distribution is as follows:

$$\sigma = \sqrt{npq} = \sqrt{(5)(0.3)(0.7)} = 1.02 \text{ papers}$$

Well, that wraps up our discussion of the binomial probability distribution. Don't be too sad, though. We'll have a chance to see this again in future chapters.

Your Turn

1. What is the probability of seeing exactly 7 heads after tossing a coin 10 times?

2. Goldey-Beacom College accepts 75 percent of applications that are submitted for entrance. What is the probability that exactly 3 of the next 6 applications will be accepted?

3. Michael Jordan makes 80 percent of his free throws. What is the probability that he will make at least 6 of his next 8 free-throw attempts?

4. A student randomly guesses at a 12-question multiple-choice test where each question has 5 choices. What is the probability that the student will correctly answer exactly 6 questions?

5. Historical records show that 5 percent of people who visit a particular website purchase something. What is the probability that no more than 2 people out of the next 7 will purchase something?

6. During the 2002 Major League Baseball season, Barry Bonds had a 0.370 batting average. Construct a binomial probability distribution for the number of successes (hits) for 5 official at bats during this season.

The Least You Need to Know

- ◆ A binomial experiment has only 2 possible outcomes for each trial.

- ◆ For a binomial experiment, the probability of success and failure is constant.

- ◆ Each trial of a binomial experiment is independent of any other trial in the experiment.

- ◆ The probability of r successes in n trials using the binomial distribution is as follows: $P[r, n] = \dfrac{n!}{(n-r)!\,r!} p^r q^{n-r}$.

- ◆ The mean for a binomial probability distribution can be calculated using the equation $\mu = np$.

- ◆ The standard deviation for a binomial probability distribution can be calculated using the equation $\sigma = \sqrt{npq}$.

The Poisson Probability Distribution

In This Chapter

- ◆ Describe the characteristics of a Poisson process
- ◆ Calculating probabilities using the Poisson equation
- ◆ Using the Poisson probability tables
- ◆ Using Excel to calculate Poisson probabilities
- ◆ Using the Poisson equation to approximate the binomial equation

Now that we have mastered the binomial probability distribution, we are ready to move on to the next discrete theoretical distribution, the Poisson. This probability distribution is named after Simeon Poisson, a French mathematician who developed the distribution during the early 1800s.

The Poisson distribution is useful for calculating the probability that a certain number of events will occur over a specific period of time. This distribution could be used to determine the likelihood that 10 customers will walk into a store during the next hour or that 2 car accidents will occur at a busy intersection this month. So let's grab some crêpes and croissants and learn about some French math.

Characteristics of a Poisson Process

In Chapter 9, we defined a binomial experiment, otherwise known as a Bernoulli process, as counting the number of successes over a specific number of trials. The result of each trial is either a success or a failure. A *Poisson process* counts the number of occurrences of an event over a period of time, area, distance, or any other type of measurement.

Rather than being limited to only 2 outcomes, the Poisson process can have any number of outcomes over the unit of measurement. For instance, the number of customers who walk into our local convenience store during the next hour could be 0, 1, 2, 3, or so on. The random variable for the Poisson distribution would be the actual number of occurrences—in this case, the number of customers arriving during the next hour.

The mean for a Poisson distribution is the average number of occurrences that would be expected over the unit of measurement. For a Poisson process, the mean has to be the same for each interval of measurement. For instance, if the average number of customers walking in the store each hour is 11, this average needs to apply to every 1-hour increment.

The last characteristic of a Poisson process is that the number of occurrences during 1 interval is independent of the number of occurrences in other intervals. In other words, if 6 customers walk in the store during the first hour of business, this will have no effect on the number of customers arriving during the second hour.

 Stat Facts _____

A **Poisson process** has the following characteristics: (1) The experiment consists of counting the number of occurrences of an event over a period of time, area, distance, or any other type of measurement; (2) the mean of the Poisson distribution has to be the same for each interval of measurement; (3) the number of occurrences during 1 interval is independent of the number of occurrences in any other interval.

Examples of random variables that may follow a Poisson probability distribution include the following:

♦ The number of cars that arrive at a tollbooth over a specific period of time

♦ The number of typographical errors found in a manuscript

♦ The number of students who are absent in my Monday-morning statistics class

♦ The number of professional football players who are placed on the injured list each week

Now that you understand the basics of a Poisson process, we are ready to move into probability calculations.

The Poisson Probability Distribution

If a random variable follows a pattern consistent with a Poisson probability distribution, we can calculate the probability of a certain number of occurrences over a given interval. To make this calculation, we need to know the average number of occurrences for the event over this interval. To demonstrate the use of the Poisson probability distribution, I'll use the following example.

The following story is true. The names have not been changed because nobody in this story is innocent. Not that any of the previous stories have been false, but this one is "really" true. Each year, Brian, John, and I make a golf pilgrimage to Myrtle Beach, South Carolina. On our last night one particular year, we were browsing through a golf store. Brian somehow convinced me to purchase a used, fancy, brand-name golf club that he swore he absolutely had to have in order to reach his full potential as a golfer. Even used, this club cost more than any that I had purchased new. Teenagers have this special talent that allows them to disregard any rational adult logic when their minds are made up.

Early the following morning, we packed our bags, checked out of the hotel, and drove to our final round of golf, which I cleverly planned to be along our route back home. On the first tee, Brian pulls out his new, used prize possession and proceeds to hit a "duck hook," which is a golfer's term for a ball that goes very short and very left, often into a bunch of trees never to be seen again. I smile nervously at Brian and try to convince myself that he'll be fine on the next hole. After hitting duck hooks on holes 2, 3, and 4, I find myself physically restraining Brian from throwing his new, used prized possession into the lake.

After our round is over, I proceed *back* to Myrtle Beach to return the club, adding an hour to what would have been a 10-hour car ride. I just hope Brian remembers times like these when I'm a frail old man drooling away in a retirement home. At the golf store, the woman cheerfully says they will take the club back, but I need to show her … *the receipt*. Now I vaguely remember putting the receipt someplace "special" just in case I would need it, but after packing, checking out, and playing golf, I would have a better chance of discovering a cure for cancer than remembering where I put that piece of paper.

Not being one to give up so easily, I march back to the car and start unpacking everything. After a short while, during which I have spread out my dirty underwear and socks all over their parking lot, the same woman walks out to tell me the store would gladly refund my money *without the receipt* if I would just pack up my things and put them back in the car.

I have discovered a very powerful technique here that I am going to pass along to you. Just consider this a bonus for using my book. Whenever I can't find a receipt when I need to return something, I simply take along some dirty clothes in a suitcase and reenact my Myrtle Beach scenario right in front of the store. Works like a charm.

Anyway, let's assume that the number of tee shots that Brian normally hits that actually land in the fairway during a round of golf is 5. The fairway is the area of short grass where the people who have designed this nerve-wracking game intended your tee shot to land. We will also assume that the actual number of fairways that Brian "hits" during 1 round follows the Poisson distribution.

Wrong Number

How do I know that the actual number of fairways that Brian "hits" during 1 round follows the Poisson distribution? At this point, I really don't know for sure. What I would need to do to verify this claim is to record the number of fairways hit over several rounds and then perform a "Goodness of Fit test" to decide whether the data fits the pattern of a Poisson distribution. I promise you that we will perform this test in Chapter 18, so please be patient.

We can now use the Poisson probability distribution to calculate the probability that Brian will hit x number of fairways during his next round, as follows:

$$P[x] = \frac{\mu^x \cdot e^{-\mu}}{x!}$$

where:

x = the number of occurrences of interest over the interval

μ = mu, the mean number of occurrences over the interval

e = the mathematical constant 2.71828

$P[x]$ = the probability of exactly x occurrences over the interval

We can now calculate the probability that Brian will hit exactly 7 fairways during his next round. With $\mu = 5$, the equation becomes this:

$$P[7] = \frac{(5^7)(2.71838^{-5})}{7!}$$

$$P[7] = \frac{(78125)(0.006738)}{7*6*5*4*3*2*1} = 0.1044$$

In other words, Brian has slightly more than a 10 percent chance of hitting exactly 7 fairways.

We can also calculate the cumulative probability that Brian will hit no more than 2 fairways using the following equations:

$$P[x \leq 2] = P[x = 0] + P[x = 1] + P[x = 2]$$

$$P[x = 0] = \frac{\left(5^0\right)\left(2.71838^{-5}\right)}{0!} = \frac{(1)(0.006738)}{1} = 0.0067$$

$$P[x = 1] = \frac{\left(5^1\right)\left(2.71838^{-5}\right)}{1!} = \frac{(5)(0.006738)}{1} = 0.0337$$

$$P[x = 2] = \frac{\left(5^2\right)\left(2.71838^{-5}\right)}{2!} = \frac{(25)(0.006738)}{2 \cdot 1} = 0.0842$$

$$P[x \leq 2] = 0.0067 + 0.0337 + 0.0842 = 0.1246$$

There is a 12.46 percent chance that Brian will hit no more than 2 fairways during his next round.

In the previous example, the mean of the Poisson distribution happened to be an integer (5). However, this doesn't have to always be the case. Suppose the number of absent students for my Monday-morning statistics follows a Poisson distribution with the average being 2.4 students. The probability that there will be 3 students absent next Monday is as follows:

Bob's Basics

Some statistics books use the symbol λ, pronounced lambda, to denote the mean of a Poisson probability distribution. However, regardless of the notation, it's still the same equation.

Bob's Basics

Remember from Chapter 8, $0! = 1$. Also $x^0 = 1$ for any value of x.

$$P[x = 3] = \frac{\left(2.4^3\right)\left(2.71838^{-2.4}\right)}{3!}$$

$$P[x = 3] = \frac{(13.824)(0.090718)}{3 * 2 * 1} = 0.2090$$

Looks like I need to start taking roll on Mondays!

There's one more cool feature of the Poisson distribution: The variance of the distribution is the same as the mean. In other words:

$$\sigma^2 = \mu$$

This means that there are no nasty variance calculations that we dealt with in previous chapters for this distribution.

Poisson Probability Tables

Just like the binomial distribution, the Poisson probability distribution has a table that allows you to look up the probabilities for certain mean values. You can find the Poisson distribution table in Appendix B of this book. The following is an excerpt from this appendix with the probabilities from our Myrtle Beach example underlined.

Values of μ

x	3.2	3.4	3.6	3.8	4.0	4.2	4.4	4.6	4.8	5.0
0	0.0408	0.0334	0.0273	0.0224	0.0183	0.0150	0.0123	0.0101	0.0082	0.0067
1	0.1304	0.1135	0.0984	0.0850	0.0733	0.0630	0.0540	0.0462	0.0395	0.0337
2	0.2087	0.1929	0.1771	0.1615	0.1465	0.1323	0.1188	0.1063	0.0948	0.0842
3	0.2226	0.2186	0.2125	0.2046	0.1954	0.1852	0.1743	0.1631	0.1517	0.1404
4	0.1781	0.1858	0.1912	0.1944	0.1954	0.1944	0.1917	0.1875	0.1820	0.1755
5	0.1140	0.1264	0.1377	0.1477	0.1563	0.1633	0.1687	0.1725	0.1747	0.1755
6	0.0608	0.0716	0.0826	0.0936	0.1042	0.1143	0.1237	0.1323	0.1398	0.1462
7	0.0278	0.0348	0.0425	0.0508	0.0595	0.0686	0.0778	0.0869	0.0959	0.1044
8	0.0111	0.0148	0.0191	0.0241	0.0298	0.0360	0.0428	0.0500	0.0575	0.0653
9	0.0040	0.0056	0.0076	0.0102	0.0132	0.0168	0.0209	0.0255	0.0307	0.0363
10	0.0013	0.0019	0.0028	0.0039	0.0053	0.0071	0.0092	0.0118	0.0147	0.0181
11	0.0004	0.0006	0.0009	0.0013	0.0019	0.0027	0.0037	0.0049	0.0064	0.0082
12	0.0001	0.0002	0.0003	0.0004	0.0006	0.0009	0.0013	0.0019	0.0026	0.0034
13	0.0000	0.0000	0.0001	0.0001	0.0002	0.0003	0.0005	0.0007	0.0009	0.0013
14	0.0000	0.0000	0.0000	0.0000	0.0001	0.0001	0.0001	0.0002	0.0003	0.0005
15	0.0000	0.0000	0.0000	0.0000	0.0000	0.0000	0.0000	0.0001	0.0001	0.0002

The probability table is organized by values of μ, the average number of occurrences. Notice that the sum of each block of probabilities for a particular value of μ adds to 1.

As with the binomial tables, one limitation of using the Poisson tables is that you are restricted to using only the values of μ that are shown in the table. For instance, the previous table would not be useful for $\mu = 0.45$. Other statistics books might contain Poisson tables that are more extensive than the one in Appendix B.

The Poisson distribution for $\mu = 5$ is shown graphically in the following histogram. The probabilities in Figure 10.1 are taken from the last column in the previous table.

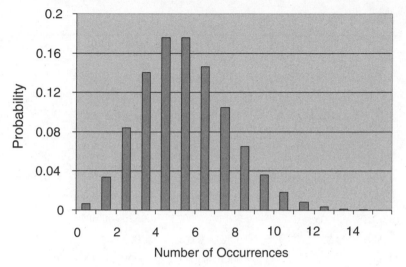

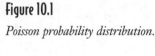

Figure 10.1

Poisson probability distribution.

Note that the most likely number of occurrences for this distribution is 4 and 5.

Here's another example. Let's assume that the number of car accidents each month at a busy intersection that I pass on my way to work follows the Poisson distribution with a mean of 1.8 accidents per month. What is the probability that there will be 3 or more accidents next month? This can be expressed as follows:

$$P[x \geq 3] = P[x = 3] + P[x = 4] + P[x = 5] + P[x = 6] + \cdots + P[x = \infty]$$

Technically, with a Poisson distribution, there is no upper limit to the number of occurrences during the interval. You'll notice from the Poisson tables that the probability of a large number of occurrences is practically 0. Because we cannot add all the probabilities of an infinite number of occurrences (if you can, you're a much better statistician than I am!), we need to take 1 minus the complement of $P[x \geq 3]$ or:

$$P[x \geq 3] = 1 - P[x < 3]$$

because:

$$P[x = 0] + P[x = 1] + P[x = 2] + P[x = 3] + P[x = 4] + P[x = 5] + \cdots + P[x = \infty] = 1.0$$

Therefore, to find the probability of 3 or more accidents, we'll use the following:

$$P[x \geq 3] = 1 - (P[x = 0] + P[x = 1] + P[x = 2])$$

Using the probabilities underlined in the following Poisson table (I seem to have misplaced my calculator), we have this:

Values of μ

x	1.1	1.2	1.3	1.4	1.5	1.6	1.7	1.8	1.9	2.0
0	0.3329	0.3012	0.2725	0.2466	0.2231	0.2019	0.1827	<u>0.1653</u>	0.1496	0.1353
1	0.3662	0.3614	0.3543	0.3452	0.3347	0.3230	0.3106	<u>0.2975</u>	0.2842	0.2707
2	0.2014	0.2169	0.2303	0.2417	0.2510	0.2584	0.2640	<u>0.2678</u>	0.2700	0.2707
3	0.0738	0.0867	0.0998	0.1128	0.1255	0.1378	0.1496	0.1607	0.1710	0.1804
4	0.0203	0.0260	0.0324	0.0395	0.0471	0.0551	0.0636	0.0723	0.0812	0.0902
5	0.0045	0.0062	0.0084	0.0111	0.0141	0.0176	0.0216	0.0260	0.0309	0.0361
6	0.0008	0.0012	0.0018	0.0026	0.0035	0.0047	0.0061	0.0078	0.0098	0.0120
7	0.0001	0.0002	0.0003	0.0005	0.0008	0.0011	0.0015	0.0020	0.0027	0.0034
8	0.0000	0.0000	0.0001	0.0001	0.0001	0.0002	0.0003	0.0005	0.0006	0.0009
9	0.0000	0.0000	0.0000	0.0000	0.0000	0.0000	0.0001	0.0001	0.0001	0.0002

$$P[x \geq 3] = 1 - (0.1653 + 0.2975 + 0.2678)$$

$$P[x \geq 3] = 1 - 0.7306 = 0.2694$$

There is almost a 27 percent chance that there will be 3 or more accidents in this intersection next month. Looks like I better find a safer way to work!

Using Excel to Calculate Poisson Probabilities

Poisson probabilities can also be conveniently calculated using Excel. The built-in POISSON function has the following characteristics:

POISSON(x, μ, cumulative)

where:

cumulative = FALSE if you want the probability of exactly x occurrences

cumulative = TRUE if you want the probability of x or fewer occurrences

For instance, Figure 10.2 shows the POISSON function being used to calculate the probability that there will be exactly 2 accidents in the intersection next month.

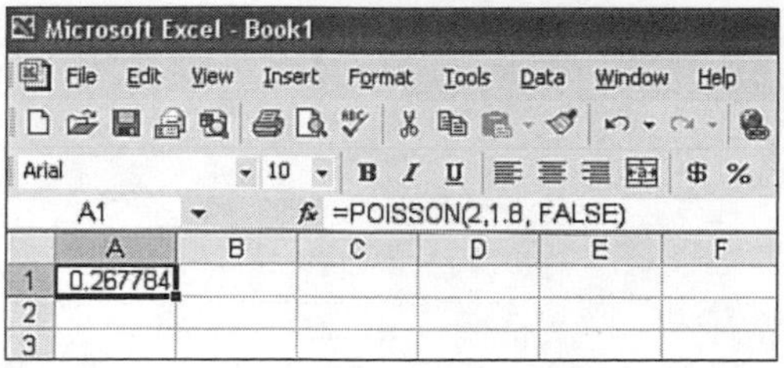

Figure 10.2

POISSON function in Excel for exactly x occurrences.

Cell A1 contains the Excel formula =POISSON(2,1.8,FALSE) with the result being 0.2678. This probability is underlined in the previous table.

Excel will also calculate the cumulative probability that there will be no more than 2 accidents in the intersection, as shown in Figure 10.3.

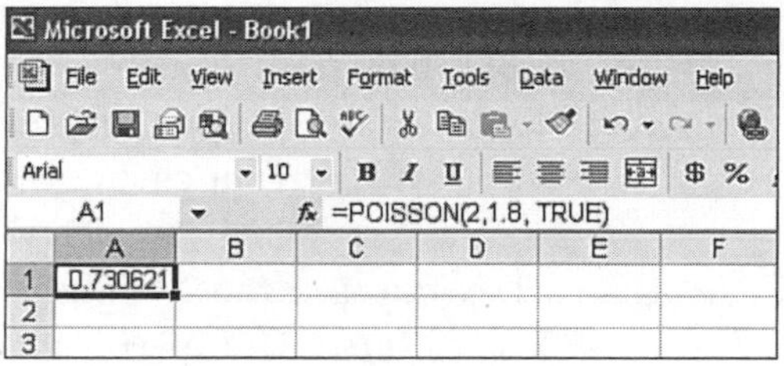

Figure 10.3

POISSON function in Excel for no more than x occurrences.

Cell A1 contains the Excel formula =POISSON(2,1.8,TRUE) with the result being 0.7306, a probability that we saw in the last calculation and which is also the sum of the underlined probabilities in the previous table.

One benefit of using Excel to determine Poisson probabilities is that you are not limited to the values of μ shown in the Poisson table in Appendix B. Excel's POISSON function allows you to use any value for μ.

Using the Poisson Distribution as an Approximation to the Binomial Distribution

I don't know about you, but when I have two ways to do something, I like to choose the one that's less work. If you don't agree with me, feel free to skip this section. If you do, read on by all means!

We can use the Poisson distribution to calculate binomial probabilities under the following conditions:

- ◆ When the number of trials, n, is greater than or equal to 20 and …

- ◆ When the probability of a success, p, less than or equal to 0.05 …

The Poisson formula would look like this:

$$P[x] = \frac{(np)^x * e^{-(np)}}{x!}$$

where:

n = the number of trials

p = the probability of a success

You might be asking yourself at this moment why you would want to do this. The answer is because the Poisson formula has fewer computations than the binomial formula and, under the stated conditions, the distributions are very close to one another.

Bob's Basics

If you need to calculate binomial probabilities with the number of trials, n, greater than or equal to 20 and the probability of a success, p, less than or equal to 0.05, you can use the equation for the Poisson distribution to approximate the binomial probabilities.

Just in case you are from Missouri (the "Show Me" state), I'll demonstrate this point with an example. Suppose there are 20 traffic lights in my town and each has a 3 percent chance of not working properly (a success) on any given day. What is the probability that exactly 1 of the 20 lights will not work today? This is a binomial experiment with n = 20, r = 1, and p = 0.03. From Chapter 9, we know that the binomial probability is this:

$$P[r, n] = \frac{n!}{(n-r)!r!} p^r q^{n-r}$$

$$P[1, 20] = \frac{20!}{(20-1)!1!}(0.03)^1(0.97)^{20-1}$$

$$P[1, 20] = (20)(0.03)(0.560613) = 0.3364$$

The Poisson approximation is as follows:

$$P[x] = \frac{(np)^x * e^{-(np)}}{x!}$$

Because $np = (20)(0.03) = 0.6$:

$$P[1] = \frac{(0.6)^1 * e^{-(0.6)}}{1!}$$

$$P[1] = (0.6)(0.548812) = 0.3293$$

Even if you're from Missouri, I think you would have to agree that the Poisson calculation is easier and the two results are very close. But if you need further proof ... Figures 10.4 and 10.5 show the histogram for each distribution for this example.

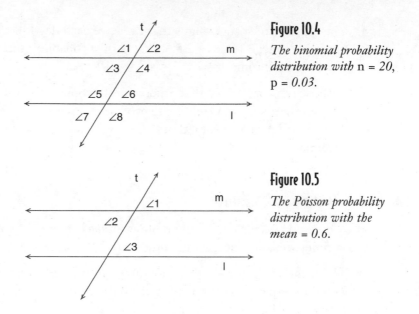

Figure 10.4

The binomial probability distribution with n = 20, p = *0.03.*

Figure 10.5

The Poisson probability distribution with the mean = 0.6.

Even to a skeptic, these two distributions look very much alike. So my advice to you is to use the Poisson equation if you're faced with calculating binomial probabilities with $n \geq 20$ and $p \leq 0.05$.

This concludes our discussion of discrete probability distributions. I hope you've had as much fun with these as I've had!

Your Turn

1. The number of rainy days per month at a particular town follows a Poisson distribution with a mean value of 6 days. What is the probability that it will rain 4 days next month?

2. The number of customers arriving at a particular store follows a Poisson distribution with a mean value of 7.5 customers per hour. What is the probability that 5 customers will arrive during the next hour?

3. The number of pieces of mail that I receive daily follows a Poisson distribution with a mean value of 4.2 per day. What is the probability that I will receive more than 2 pieces of mail tomorrow?

4. The number of employees who call in sick on Monday follows a Poisson distribution with a mean value of 3.6. What is the probability that no more than 3 employees will call in sick next Monday?

5. The number of spam e-mails that I receive each day follows a Poisson distribution with a mean value of 2.5. What is the probability that I will receive exactly 1 spam e-mail tomorrow?

6. Historical records show that 5 percent of people who visit a particular website purchase something. What is the probability that exactly 2 people out of the next 25 will purchase something? Use the Poisson distribution to estimate this binomial probability.

The Least You Need to Know

◆ A Poisson process counts the number of occurrences of an event over a period of time, area, distance, or any other type of measurement.

◆ The mean for a Poisson distribution is the average number of occurrences that would be expected over the unit of measurement and has to be the same for each interval of measurement.

◆ The number of occurrences during one interval of a Poisson process is independent of the number of occurrences in other intervals.

◆ If x is a Poisson random variable, the probability of x occurrences over the interval of measurement is $P[x] = \dfrac{\mu^x \cdot e^{-\mu}}{x!}$.

◆ If the number of binomial trials is greater than or equal to 20 and the probability of a success is less than or equal to 0.05, you can use the equation for the Poisson distribution to approximate the binomial probabilities.

The Normal Probability Distribution

In This Chapter

- ◆ Examining the properties of a normal probability distribution
- ◆ Using the standard normal table to calculate probabilities of a normal random variable
- ◆ Using Excel to calculate normal probabilities
- ◆ Using the normal distribution as an approximation to the binomial distribution

Now that we have completed our journey through discrete probability distributions, we are ready to take on a new challenge. Our next destination is continuous random variables and a continuous probability distribution known as the normal distribution. If you can remember from Chapter 8, we defined a continuous random variable as one that can assume any numerical value within an interval as a result of measuring the outcome of an experiment. Some examples of continuous random variables are weight, distance, speed, or time.

The normal distribution is a statistician's workhorse. This distribution is the foundation for many types of inferential statistics that we rely on today. We will continue to refer to this distribution for many of the remaining chapters in this book.

Characteristics of the Normal Probability Distribution

A continuous random variable that follows the normal probability distribution has several distinctive features. Let's say the monthly rainfall in inches for a particular city follows the normal distribution with an average of 3.5 inches and a standard deviation of 0.8 inches. The probability distribution for such a random variable is shown in Figure 11.1.

Figure 11.1

Normal distribution with a mean = 3.5, standard deviation = 0.8.

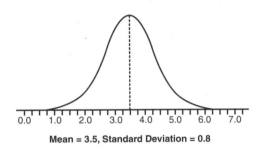

Normal Probability Distribution

Mean = 3.5, Standard Deviation = 0.8

From this figure, we can make the following observations about the normal distribution:

◆ The mean, median, and mode are the same value—in this case, 3.5 inches.

◆ The distribution is bell shaped and symmetrical around the mean.

◆ The total area under the curve is equal to 1.

◆ The left and right tails of the normal probability distribution extend indefinitely, never quite touching the horizontal axis.

The standard deviation plays an important role in the shape of the curve. Looking at the previous figure, nearly all the monthly rainfall measurements would fall between 1.0 and 6.0 inches. Now look at Figure 11.2, which shows the normal distribution with the same mean of 3.5 inches, but with a standard deviation of only 0.5 inches.

Normal Probability Distribution

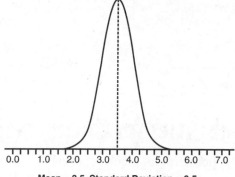

Mean = 3.5, Standard Deviation = 0.5

Figure 11.2

Normal distribution with a mean = 3.5, standard deviation = 0.5.

Here you can see a curve that's much tighter around the mean. Almost all the rainfall measurements will be between 2.0 and 5.0 inches per month.

Figure 11.3 shows the impact of changing the mean of the distribution to 5.0 inches, leaving the standard deviation at 0.8 inches.

Bob's Basics

A smaller standard deviation results in a "skinnier" curve that's tighter and taller around the mean. A larger σ (standard deviation) makes for a "fatter" curve that's more spread out and not as tall.

Normal Probability Distribution

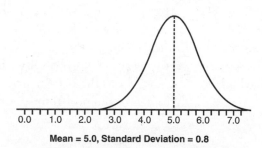

Mean = 5.0, Standard Deviation = 0.8

Figure 11.3

Normal distribution with a mean = 5.0, standard deviation = 0.8.

In each of the previous figures, the characteristics of the normal probability distribution hold true. In each case, the values of μ, the mean, and σ, the standard deviation, completely describe the shape of the distribution.

The probability function for the normal distribution has a particularly mean personality (that pun was surely intended) and is shown as follows:

$$f[x] = \frac{1}{\sigma\sqrt{2\pi}} e^{-(1/2)[(x-\mu)/\sigma]^2}$$

I promise you this will be the last you'll see of this beast. Fortunately, we have other methods for calculating probabilities for this distribution that are more civilized and are discussed starting in the next section.

Calculating Probabilities for the Normal Distribution

There are a couple of approaches to calculate probabilities for a normal random variable. I'll use the following example to demonstrate how this is done.

One morning a few days ago, Deb called me on my cell phone while I was out running errands and spoke the two words that I had feared hearing for the past year. "They're back," she said. "Okay," I replied somberly, and then hung up the phone and headed straight toward the hardware store. My manhood was once again being challenged and I'd be darned if I was going to take this lying down. This was *war* and I was coming home fully prepared for battle. I am referring to, of course, my annual struggle with the most vile, the most dastardly, the most hungry creature that God has ever placed on this planet … the *Japanese beetle*.

By the time I returned home from the hardware store, half of our beautiful plum tree looked like Swiss cheese. I quickly counterattacked with a vengeance, spraying the most potent chemicals money could buy. In the end, after the toxic spray clears, I stand alone, master of my domain.

Alright, let's say that the amount of toxic spray I use each year follows a normal distribution with a mean of 60 ounces and a standard deviation of 5 ounces. This means that each year that I do battle with these demons, the most likely amount of spray I'll use is 60 ounces, but it will vary year to year. The probability of other amounts above and below 60 ounces will drop off according to the bell-shaped curve. Armed with this information, we are now ready to determine probabilities of various usages each year.

Calculating the Standard Z-Score

Because the total area under a normal distribution curve equals 1 and the curve is symmetrical, we can say the probability that I will use 60 ounces or more of spray is 50 percent, as is the probability that I will use 60 ounces or less. This is shown in Figure 11.4.

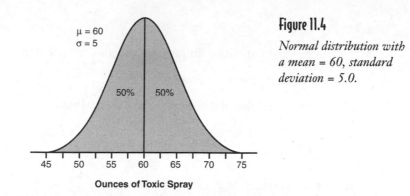

Figure 11.4

Normal distribution with a mean = 60, standard deviation = 5.0.

How would you calculate the probability that I would use 64.3 ounces of spray or less next year? I'm glad you asked. For this task, we need to define the *standard normal* distribution, which is a normal distribution with a $\mu = 0$ and $\sigma = 1.0$, and is shown in Figure 11.5.

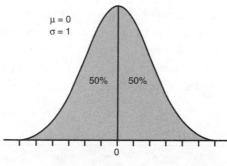

Figure 11.5

Standard normal distribution with a mean = 0, standard deviation = 1.0.

This standard normal distribution is the basis for all normal probability calculations and is used throughout this chapter.

My next step is to determine how many standard deviations the value 64.3 is from the mean of 60 and show this value on the standard normal distribution curve. We do this using the following formula:

$$z = \frac{x - \mu}{\sigma}$$

Stat Facts

The **standard normal** distribution is a normal distribution with a mean equal to 0 and a standard deviation equal to 1.0.

where:

x = the normally distributed random variable of interest

μ = the mean of the normal distribution

σ = the standard deviation of the normal distribution

z = the number of standard deviations between x and μ, otherwise known as *the standard z-score*.

For the current example, the standard z-score is as follows:

$$z_{64.3} = \frac{64.3 - 60}{5} = 0.86$$

Now I know that 64.3 is 0.86 standard deviations away from 60 in my distribution.

Using the Standard Normal Table

Now that I have my standard z-score, I can use the following table to determine the probability that I will use 64.3 ounces of toxic spray or less next year. This table is an excerpt from Appendix B and shows the area of the standard normal curve up to and including certain values of z. Because $z = 0.86$ in this example, we go to the 0.8 row and the 0.06 column to find a value of 0.8051, which is underlined.

Second Digit of Z

z	0.00	0.01	0.02	0.03	0.04	0.05	0.06	0.07	0.08	0.09
0.0	0.5000	0.5040	0.5080	0.5120	0.5160	0.5199	0.5239	0.5279	0.5319	0.5359
0.1	0.5398	0.5438	0.5478	0.5517	0.5557	0.5596	0.5636	0.5675	0.5714	0.5753
0.2	0.5793	0.5832	0.5871	0.5910	0.5948	0.5987	0.6026	0.6064	0.6103	0.6141
0.3	0.6179	0.6217	0.6255	0.6293	0.6331	0.6368	0.6406	0.6443	0.6480	0.6517
0.4	0.6554	0.6591	0.6628	0.6664	0.6700	0.6736	0.6772	0.6808	0.6844	0.6879
0.5	0.6915	0.6950	0.6985	0.7019	0.7054	0.7088	0.7123	0.7157	0.7190	0.7224
0.6	0.7257	0.7291	0.7324	0.7357	0.7389	0.7422	0.7454	0.7486	0.7517	0.7549
0.7	0.7580	0.7611	0.7642	0.7673	0.7704	0.7734	0.7764	0.7794	0.7823	0.7852
0.8	0.7881	0.7910	0.7939	0.7967	0.7995	0.8023	_0.8051_	0.8078	0.8106	0.8133
0.9	0.8159	0.8186	0.8212	0.8238	0.8264	0.8289	0.8315	0.8340	0.8365	0.8389
1.0	0.8413	0.8438	0.8461	0.8485	0.8508	0.8531	0.8554	0.8577	0.8599	0.8621

This area is shown graphically in Figure 11.6.

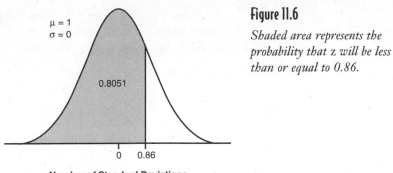

Figure 11.6

Shaded area represents the probability that z will be less than or equal to 0.86.

The probability that the standard z-score will be less than or equal to 0.86 is 80.51 percent. Because:

$$P[z \leq 0.86] = P[x \leq 64.3] = 0.8051$$

There is an 80.51 percent chance I will use 64.3 ounces of spray or less next year against those evil Japanese beetles. This can be seen in Figure 11.7.

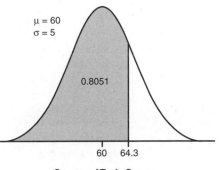

Figure 11.7

Shaded area represents the probability that x will be less than or equal to 64.3 ounces.

CAUTION

Wrong Number

With continuous random variables, we cannot determine the probability of using exactly 64.3 ounces of spray because this would be an infinitely small probability. This is because there are an infinite amount of quantities that I can use any given year. One year, I could use 61.757 ounces and another year, 53.472 ounces. That's why with continuous random variables we can only calculate the *probabilities* of certain intervals, like less than 64.3 ounces or between 50.5 and 58.1 ounces. Compare this to discrete random variables from previous chapters. Because there were only a finite number of values for these variables, we could calculate the probability of exactly *x* occurrences or *r* successes.

What about the probability that I will use more than 62.5 ounces of spray next year? Because the standard normal table only has probabilities that are less than or equal to the z-scores, we need to look at the complement to this event.

$$P[x > 62.5] = 1 - P[x \leq 62.5]$$

The z-score now becomes this:

$$z_{62.5} - \frac{62.5 - 60}{5} = 0.50$$

According to our normal table:

$$P[z \leq 0.50] = 0.6915$$

But we want:

$$P[z > 0.50] = 1 - 0.6915 = 0.3085$$

This probability is shown graphically in Figure 11.8.

Figure 11.8

Shaded area represents the probability that z will be more than 0.50.

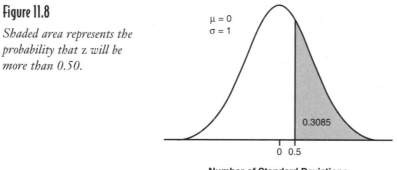

Because:

$$P[z > 0.50] = P[x > 62.5] = 0.3085$$

There is a 30.85 percent chance that I will use more than 62.5 ounces of toxic spray. Beetles beware!

What about the probability that I will use more than 54 ounces of spray? Again, I need the complement, which would be this:

$$P[x > 54] = 1 - P[x \leq 54]$$

The z-score becomes this:

$$z_{54} = \frac{54 - 60}{5} = -1.20$$

The negative score indicates that we are to the left of the distribution mean. Notice that the standard normal table only shows positive z values. But this is no problem because the distribution is symmetric. Figure 11.9 shows that the shaded area to the left of –1.2 standard deviations from the mean is the same as the shaded area to the right of +1.2 standard deviations from the mean.

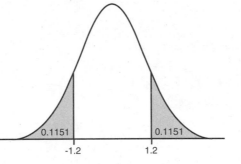

Figure 11.9

The shaded areas are equal.

0.1151   0.1151

-1.2   1.2

Number of Standard Deviations

We can determine the area to the right of +1.2 standard deviations as follows:

$$P[z > +1.2] = 1 - P[z \le +1.2] = 1 - 0.8849 = 0.1151$$

Therefore, the area to the left of –1.2 standard deviations from the mean is also 0.1151. We now can calculate the area to the right of –1.2 standard deviations from the mean.

$$P[z > -1.2] = 1 - P[z \le -1.2] = 1 - 0.1151 = 0.8849$$

Because:

$$P[x > 54] = P[z > -1.2] = 0.8849$$

There is an 88.49 percent chance I will use more than 54 ounces of spray. This probability is shown graphically in Figure 11.10.

Figure 11.10

The shaded area is the probability that x will be more than 54 ounces.

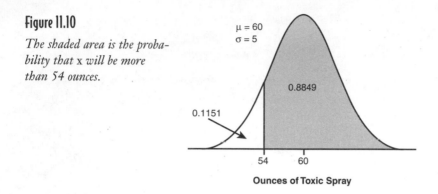

Bob's Basics

A shortcut to the previous example would be to recognize the following:

$$P[z > -1.20] = P[z \leq +1.20]$$

$$P[z > -1.20] = 0.8849$$

In general, you can use the following two relationships for any value a when dealing with negative z-scores:

$$P[z > -a] = P[z \leq +a]$$

$$P[z \leq -a] = 1 - P[z \leq +a]$$

Finally, let's look at the probability that I will use between 54 and 62.5 ounces of spray next year. This probability is shown graphically in Figure 11.11.

Figure 11.11

The shaded area is the probability that x will be between 54 and 62.5 ounces.

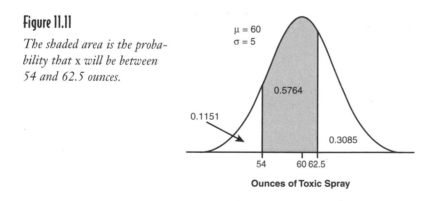

We know from previous examples that the area to the left of 54 ounces is 0.1151 and that the area to the right of 62.5 ounces is 0.3085. Because the total area under the curve is 1:

$$P[54 \leq x \leq 62.5] = 1 - 0.1151 - 0.3085 = 0.5764$$

There is a 57.64 percent chance that I will use between 54 and 62.5 ounces of spray next year. I can't wait.

The Empirical Rule Revisited

Remember way back in Chapter 5 we discussed the empirical rule, which stated that if a distribution follows a bell-shaped, symmetrical curve centered around the mean, we would expect approximately 68, 95, and 99.7 percent of the values to fall within 1.0, 2.0, and 3.0 standard deviations around the mean respectively. I'm glad to inform you that we now have the ability to demonstrate these results.

The shaded area in Figure 11.12 shows the percentage of observations that we would expect to fall within 1.0 standard deviation of the mean.

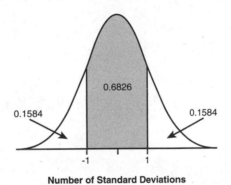

Number of Standard Deviations

Figure 11.12

The shaded area is the probability that x *will be between −1.0 and +1.0 standard deviations from the mean.*

Where did 68 percent come from? We can look in the normal table to get the probability that an observation will be less than 1 standard deviation from the mean.

$$P[z \leq +1.0] = 0.8413$$

Therefore, the area to the right of +1.0 standard deviations is this:

$$P[z > +1.0] = 1 - 0.8413 = 0.1587$$

By symmetry, the area to the left of −1.0 standard deviations is also 0.1587. That leaves the area between −1.0 and +1.0 as this:

$$P[-1.0 \leq z \leq +1.0] = 1 - 0.1587 - 0.1587 = 0.6826$$

The same logic is used to demonstrate the probabilities of 2.0 and 3.0 standard deviations from the mean. I'll leave those for you to try.

Calculating Normal Probabilities Using Excel

Once again we can rely on Excel to do some of the grunt work for us. The first built-in function is NORMDIST, which has the following characteristics:

NORMDIST(x, mean, standard_dev, cumulative)

where:

cumulative = FALSE if you want the probability mass function (we don't)

cumulative = TRUE if you want the cumulative probability (we do)

For instance, Figure 11.13 shows the NORMDIST function being used to calculate the probability that I will use less than 64.3 ounces of spray on those nasty beetles next year.

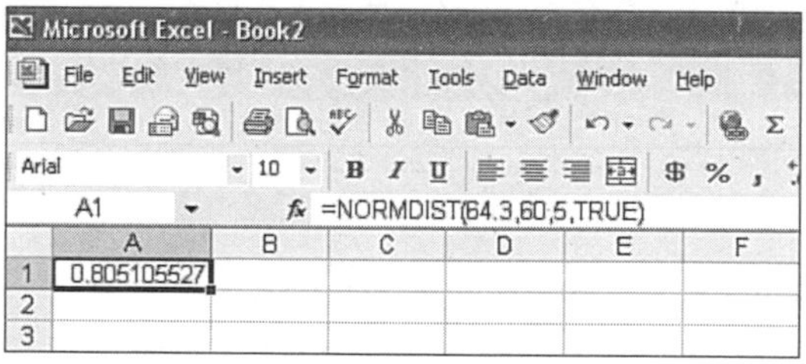

Figure 11.13

NORMDIST function in Excel for less than 64.3 ounces.

Cell A1 contains the Excel formula =NORMDIST(64.3,60,5,TRUE) with the result being 0.8051. This probability is underlined in the previous table.

Excel also has a cool function called NORMSINV, which has the following characteristics:

NORMSINV(probability)

You provide this function a probability between 0 and 1, and it returns the corresponding z-score. Figure 11.14 shows the NORMSINV function returning a z-score for a probability of 0.8413, which is 1.0 standard deviation from the mean.

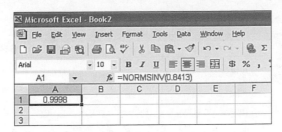

Figure 11.14

NORMSINV function in Excel for 1.0 standard deviation.

Cell A1 contains the Excel formula =NORMSINV(0.8413) with the result being 0.9998 (close enough to 1.0). If you refer back to Figure 11.12, notice that the area to the left of 1.0 standard deviation from the mean totals to 0.8413. You can also find this value in the standard normal table next to $z = 1.0$.

Using the Normal Distribution as an Approximation to the Binomial Distribution

Remember how nasty our friend the binomial distribution can get sometimes? Well, the normal distribution may be able help us out during these difficult times under the right conditions. Recall from Chapter 9, the binomial equation will calculate the probability of r successes in n trials with p = the probability of a success for each trial and q = probability of a failure. If $np \geq 5$ and $nq \geq 5$, we can use the normal distribution to approximate the binomial.

As an example, suppose my statistics class is composed of 60 percent females. If I select 15 students at random, what is the probability that this group will include 8, 9, 10, or 11 female students? For this example, $n = 15$, $p = 0.6$, $q = 0.4$, and $r =$ 8, 9, 10, and 11. We can use the normal approximation because np = (15)(0.6) = 9 and $nq = (15)(0.4) = 6$. (Sorry, guys. I didn't mean to infer picking you would be classified a failure!)

This can be solved using the binomial table in Appendix B. A portion of this table is shown with the probabilities of interest underlined.

Bob's Basics

Even if you are not interested in learning how the normal distribution can be used to approximate the binomial, I strongly encourage you to work through the example in this section. It will be good practice for determining probabilities for a normal distribution. And we all know that practice makes perfect!

Stat Facts

The normal distribution can be used to approximate the binomial distribution when $np \geq 5$ and $nq \geq 5$.

Values of *p*

n r	0.1	0.2	0.3	0.4	0.5	0.6	0.7	0.8	0.9
15 0	0.2059	0.0352	0.0047	0.0005	0.0000	0.0000	0.0000	0.0000	0.0000
1	0.3432	0.1319	0.0305	0.0047	0.0005	0.0000	0.0000	0.0000	0.0000
2	0.2669	0.2309	0.0916	0.0219	0.0032	0.0003	0.0000	0.0000	0.0000
3	0.1285	0.2501	0.1700	0.0634	0.0139	0.0016	0.0001	0.0000	0.0000
4	0.0428	0.1876	0.2186	0.1268	0.0417	0.0074	0.0006	0.0000	0.0000
5	0.0105	0.1032	0.2061	0.1859	0.0916	0.0245	0.0030	0.0001	0.0000
6	0.0019	0.0430	0.1472	0.2066	0.1527	0.0612	0.0116	0.0007	0.0000
7	0.0003	0.0138	0.0811	0.1771	0.1964	0.1181	0.0348	0.0035	0.0000
8	0.0000	0.0035	0.0348	0.1181	0.1964	0.1771	0.0811	0.0138	0.0003
9	0.0000	0.0007	0.0116	0.0612	0.1527	0.2066	0.1472	0.0430	0.0019
10	0.0000	0.0001	0.0030	0.0245	0.0916	0.1859	0.2061	0.1032	0.0105
11	0.0000	0.0000	0.0006	0.0074	0.0417	0.1268	0.2186	0.1876	0.0428

Also recall from Chapter 9 that the mean and standard deviation of this binomial distribution is this:

$$\mu = np = (15)(0.6) = 9$$

$$\sigma = \sqrt{npq} = \sqrt{(15)(0.6)(0.4)} = 1.897$$

The probability that the group of 15 students will include 8, 9, 10, or 11 female students is as follows:

$$P[r = 8,\ 9,\ 10,\ \text{or } 11] = 0.1771 + 0.2066 + 0.1859 + 0.1268 = 0.6964$$

Now let's solve this problem using the normal distribution and compare the results. Figure 11.15 shows the normal distribution with $\mu = 9$ and $\sigma = 1.897$.

Figure 11.15

The normal approximation to the binomial distribution.

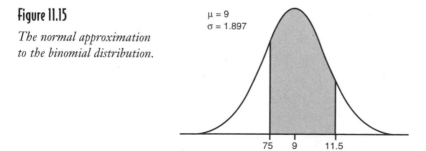

Number of Female Students

Notice that the shaded interval goes from 7.5 to 11.5 rather than 8 to 11. Don't worry, I didn't make a mistake. I subtracted 0.5 from 8 and added 0.5 to 11 to compensate for the fact that the normal distribution is continuous and the binomial is discrete. Adding and subtracting 0.5 is known as the *continuity correction factor*. For larger values of *n*, like 100 or more, you can ignore this correction factor.

Now we need to calculate the z-scores.

$$z_{11.5} = \frac{x - \mu}{\sigma} = \frac{11.5 - 9}{1.897} = +1.32$$

$$z_{7.5} = \frac{x - \mu}{\sigma} = \frac{7.5 - 9}{1.897} = -0.79$$

According to the normal table:

$$P[z \le +1.32] = 0.9066$$

This area is shown in the shaded region of Figure 11.16.

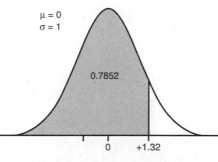

$\mu = 0$
$\sigma = 1$

0.7852

0   +1.32

Number of Standard Deviations

Figure 11.16

The probability that $z \le +1.32$ standard deviations from the mean.

We also know because of symmetry with the normal curve that:

$$P[z \le -0.79] = 1 - P[z \le +0.79]$$

According to the normal table:

$$P[z \le +0.79] = 0.7852$$

Therefore:

$$P[z \le -0.79] = 1 - 0.7852 = 0.2148$$

This probability is shown in the shaded area in Figure 11.17.

Figure 11.17

The probability that z ≤ −0.79 standard deviations from the mean.

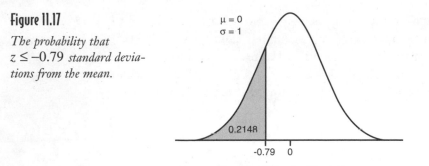

0.2148

-0.79   0

Number of Standard Deviations

The probability of interest for this example is the area between z-scores of −0.79 and +1.32. We can use the following calculations to find this area:

$$P[-0.79 \le z \le +1.32] = P[z \le +1.32] - P[z \le -0.79]$$

$$P[-0.79 \le z \le +1.32] = 0.9066 - 0.2148 = 0.6918$$

This probability is shown in the shaded area in Figure 11.18.

Figure 11.18

The probability that −0.79 ≤ z ≤ +1.32 standard deviations from the mean.

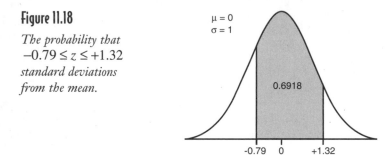

0.6918

-0.79   0   +1.32

Number of Standard Deviations

Using the normal distribution, we have determined the probability that my group of 15 students will contain 8, 9, 10, or 11 females is 0.6916. As you can see, this probability is very close to the result we obtained using the binomial tables, which was 0.6964.

Well, that ends our chapter on the normal probability distribution. I feel much better prepared for next year's return visit of my archenemy, the Japanese beetle. Wish me luck.

Your Turn

1. The speed of cars passing through a checkpoint follows a normal distribution with μ = 62.6 miles per hour and σ = 3.7 miles per hour. What is the probability that the next car passing will …

 a. Be exceeding 65.5 miles per hour?

 b. Be exceeding 58.1 miles per hour?

 c. Be between 61 and 70 miles per hour?

2. The selling price of various homes in a community follows the normal distribution with $\mu = \$176,000$ and $\sigma = \$22,300$. What is the probability that the next house will sell for ...

 a. Less than $176,000?

 b. Less than $158,000?

 c. Between $150,000 and $168,000?

3. The age of customers for a particular retail store follows a normal distribution with $\mu = 37.5$ years and $\sigma = 7.6$ years. What is the probability that the next customer who enters the store will be ...

 a. More than 31 years old?

 b. Less than 42 years old?

 c. Between 40 and 45 years old?

4. A coin is flipped 14 times. Use the normal approximation to the binomial distribution to calculate the probability of a total of 4, 5, or 6 heads. Compare this to the binomial probability.

The Least You Need to Know

 ◆ The normal distribution is bell shaped and symmetrical around the mean.

 ◆ The total area under the normal distribution curve is equal to 1.0.

 ◆ The normal distribution tables are based on the standard normal distribution where $\mu = 0$ and $\sigma = 1.0$.

 ◆ The number of standard deviations between a normally distributed random variable, x, and μ is known as the standard z-score and can be found with
 $$z = \frac{x - \mu}{\sigma}.$$

 ◆ Excel has two built-in functions that you can use to perform normal distribution calculations: NORMDIST and NORMSINV.

 ◆ The normal distribution can be used to approximate the binomial distribution when $np \geq 5$ and $nq \geq 5$.

Part 3

Inferential Statistics

Now we can take all those wonderful concepts that we have stuffed into our overloaded brains from Parts 1 and 2 and put them to work using statistically sounding words such as *confidence interval* and *hypothesis test*. Inferential statistics enables us to make statements about a general population using the results of a random sample from that population. For instance, using inferential statistics, the winner of a political election can be accurately predicted very early in the polling process based on the results of a relatively small random sample that is properly chosen. Pretty cool stuff!

Chapter 12

Sampling

In This Chapter

◆ The reason for measuring a sample rather than the population

◆ The various methods for collecting a random sample

◆ Defining sampling errors

◆ Consequences for poor sampling techniques

Our first chapter dealing with the long-awaited topic of inferential statistics focuses on the subject of sampling. If you can think way back to Chapter 1, we defined a population as representing all possible outcomes or measurements of interest, and a sample as a subset of a population. In this chapter we'll talk about why we use samples in statistics and what can go wrong if they are not used properly.

Virtually all statistical results are based on the measurements of a sample drawn from a population. Major decisions are often made based on information from samples. For instance, the Nielson ratings gather information from a small sample of homes and are used to infer the television-viewing patterns of the entire country. The future of your favorite TV show rests in the hands of these select few! Choosing the proper sample is a critical step to ensure accurate statistical conclusions.

Why Sample?

Most statistical studies are based on a sample of the population at large. The relationship between a population and sample is shown in Figure 12.1 (and also described in Chapter 1).

Figure 12.1

The relationship between a population and sample.

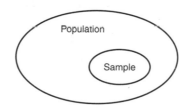

Why not just measure the whole population rather than rely on only a sample? Very good question! Depending on the study, measuring an entire population could be very expensive or just plain impossible. If I want to measure the life span of a certain breed of pesky mosquitoes (extremely short if I had any say in the matter), it's not possible to observe every single mosquito in the population. I would need to rely on a sample of the mosquito population, measure their life span, and make a statement about the life span of the entire population. That's the whole concept of inferential statistics all in one paragraph! Unfortunately, doing what I just said is a whole lot harder than just saying it. That's what the rest of this book is all about.

Even if we could feasibly measure the entire population, it often would be a wasteful decision to do so. If a sample is collected properly and the analysis is performed correctly, we can make a very accurate assessment of the entire population. There is very little added benefit to continue on beyond the sample and measure everything in sight. Measuring the population often is a waste of both time and money, resources that seem to be very scarce these days.

One example where such a decision was recently made occurred at Goldey-Beacom College, where I presently teach. I am also the Chair of the Academic Honor Code Committee and was involved in a project whose goal was to gather information regarding the attitude of our student body on the topic of academic integrity. It would have been possible to ask

> **Random Thoughts**
>
> Nielsen Media Research surveys 5,000 households nationwide to infer the television habits of millions of people. Because the results of these surveys are the basis for decisions such as show cancellations and advertising revenue, you better believe they select this sample *very* carefully.

> **Bob's Basics**
>
> It is often not feasible to measure an entire population. Even when it is feasible, measuring an entire population can be a waste of time and money and provides little added benefit beyond measuring a sample.

every student at our college to respond to the survey, but it was also unnecessary with the availability of inferential statistics. We eventually made the intelligent decision and sampled only a portion of the students to infer the attitudes of the population.

Random Sampling

The term *random sampling* refers to a sampling procedure where every member in the population has a chance of being selected. The objective of the sampling procedure is to ensure that the final sample to be measured is representative of the population from which it was taken. If this is not the case, then we have a *biased sample*, which can lead to misleading results. If you recall, we discussed an example of a biased sample back in Chapter 1 with the golf course survey. The selection of a proper sample is critical to the accuracy of the statistical analysis.

Stat Facts

Random sampling refers to a sampling procedure where every member in the population has a chance of being selected. A **biased sample** is a sample that does not represent the intended population and can lead to distorted findings.

There are several ways in which a random sample can be selected. To demonstrate these techniques, I'll use the following example.

Most of the time, I consider Deb to be a person of sound mind and judgment (she married me after all). Lately, however, I have had some concerns about her behavior dealing with the fact that she is reaching a major milestone in life before I am. Although I am not permitted to mention exactly what this milestone is (under penalty of not proofreading any more chapters and other certain activities), I will say it involves dividing the number 100 by 2. (You do the math.)

Anyway, we were recently walking through the local mall and Deb suddenly ran to a sales counter where they were selling fake ponytails for your hair. I had never in my life heard of such a thing and in a million years never would have conceived the idea. Deb, on the other hand, thought it was absolutely brilliant. Within seconds, a total stranger appeared from nowhere and before I could say, "That's my wife," had re-arranged Deb's hair and, in his final crowning moment, expertly arranged a fake hairpiece that somewhat resembled a small, furry animal on the top of her head.

Deb, beaming with her "new look," turned to me to ask what I thought. Because this also happened to be our wedding anniversary, I weakly said it looked great as I handed this total stranger my credit card. (I might be a little slow in these matters, but I'm not stupid.) Deb spent the rest of the evening prancing through the mall with her new cute furry animal hanging on for dear life. I have to admit, once I got used to the idea, it did look pretty cute.

Let's say we wanted to conduct a survey to collect opinions of Deb's new look. In fact, you, the reader, can render your opinion of Deb after observing Figure 12.2 by sending me an e-mail from the book's website at www.stat-guide.com.

Figure 12.2

Deb's new look; what do you think?

If I consider the current shoppers at the mall that night as my population, I need to decide how to select the random sample from whose opinion I will ask. As we will see in the following sections, there are four different ways to gather a random sample: simple random, systematic, cluster, and stratified.

Simple Random Sampling

A *simple random sample* is a sample in which every member of the population has an equal chance of being chosen. Unfortunately, this is easier said than done. In our mall example, I can randomly approach people to ask their opinion. However, there might

Stat Facts

A simple random sample is a sample in which every member of the population has an equal chance of being chosen.

be some biases in my selection. For instance, if I observe that a certain menacing-looking person has a tattoo that says, "Death to All Statisticians," I might choose not to ask him what he thinks of Deb's new ponytail. But in doing so, I might be biasing my sample.

Assuming I can rid myself of any biased selection, Figure 12.3 would describe a simple random sample at the mall.

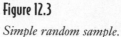

Figure 12.3

Simple random sample.

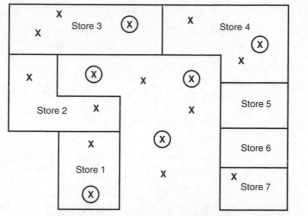

Each "X" represents a shopper, and each "X" that's circled represents a shopper in my sample.

There would be other options for choosing a simple random sample for the Academic Integrity survey mentioned earlier in the chapter. I could randomly choose students using a *random number table*, which is aptly named. (After all, it is simply a table of numbers that are completely random.) An excerpt of such a table is shown here:

57245	39666	18545	50534	57654	25519	35477	71309	12212	98911
42726	58321	59267	72742	53968	63679	54095	56563	09820	86291
82768	32694	62828	19097	09877	32093	23518	08654	64815	19894
97742	58918	33317	34192	06286	39824	74264	01941	95810	26247
48332	38634	20510	09198	56256	04431	22753	20944	95319	29515
26700	40484	28341	25428	08806	98858	04816	16317	94928	05512
66156	16407	57395	86230	47495	13908	97015	58225	82255	01956
64062	10061	01923	29260	32771	71002	58132	58646	69089	63694
24713	95591	26970	37647	26282	89759	69034	55281	64853	50837
90417	18344	22436	77006	87841	94322	45526	38145	86554	42733

Suppose there were 1,000 students in the population from which we were drawing a sample size of 100. (We'll discuss sample size in a later chapter.) I would list these students with assigned numbers from 0 to 999. The random number table would tell me to select student 572, followed by student 427, and so forth until 100 students were

selected. Using this technique, my sample of 100 students would be chosen with complete randomness.

Random numbers can also be generated with Excel using the RAND function. Figure 12.4 demonstrates how this is done.

Cell A1 contains the formula =RAND(), which provides a random number between 0 and 1. This random number would result in student 357 being chosen for the sample.

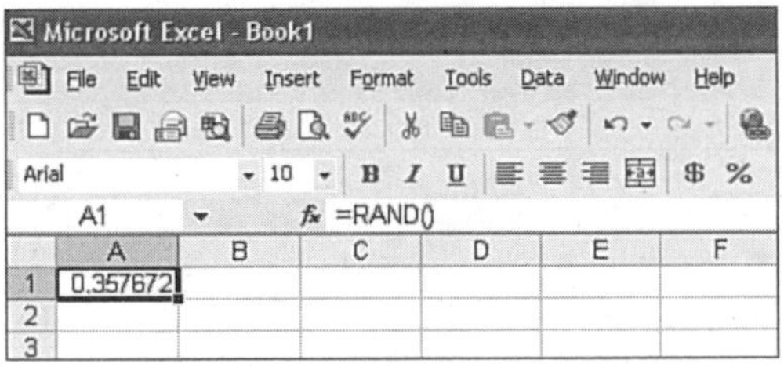

Figure 12.4

Excel's random number generator.

Systematic Sampling

One way to avoid a personal bias when selecting people at random is to use *systematic sampling.* This technique results in selecting every k^{th} member of the population to be in your sample. The value of k will depend on the size of the sample and size of the population. Using my Academic Integrity survey, with a population of 1,000 students and a sample of 100, $k = 10$. From a listing of the entire population, I would choose every tenth student to be included in the sample. In general, if N = the size of the population, and n = the size of the sample, the value of k would be approximately $\dfrac{N}{n}$.

We could also apply this sampling technique to the mall survey. Figure 12.5 shows every third customer walking into the mall being asked his or her opinion of Deb's ponytail, even if the customer does have a tattoo.

Again, each "X" represents a shopper, and each "X" that's circled represents a shopper in my sample.

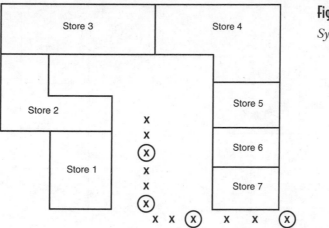

Figure 12.5
Systematic sampling.

The benefit of systematic sampling is that it's easier to conduct than a simple random sample, often resulting in less time and money. The downside is the danger of selecting a biased sample if there is a pattern in the population that is consistent with the value of *k*. For instance, let's say I'm conducting a survey on campus asking students how many hours they are studying during the week and I select every fourth week to collect my data. Because we are on an 8-week semester schedule at Goldey-Beacom, every fourth week could end up being mid-terms and finals week, which would result in a higher number of study hours than normal (or at least I would hope so!).

Cluster Sampling

If the population can be divided into groups, or *clusters*, then a simple random sample can be selected from these clusters to form the final sample. Using the Academic Integrity survey, the clusters could be defined as classes. We would randomly choose different classes to participate in the survey. In each class chosen, every student would be selected to be part of the sample.

The mall survey could also be conducted using cluster sampling. Clusters could be defined as stores in the mall population. Different stores could be randomly chosen and each customer in these stores could be asked his or her opinion about Deb's ponytail. Figure 12.6 shows cluster sampling graphically.

Stat Facts

A **cluster** sample is a simple random sample of groups, or clusters, of the population. Each member of the chosen clusters would be part of the final sample.

Figure 12.6

Cluster sampling.

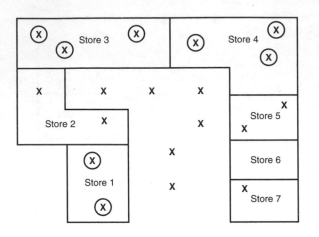

According to this figure, Stores 1, 3, and 4 have been chosen to participate in the survey.

For cluster sampling to be effective, it is assumed that each cluster that is selected for the sample is representative of the population at large. In effect, each cluster is a miniaturized version of the overall population. If used properly, cluster sampling can be a very cost-effective way of collecting a random sample from the population. In the mall example, I would only have to visit three stores to conduct my survey, saving me valuable time on my wedding anniversary.

Stratified Sampling

In *stratified sampling*, we divide the population into mutually exclusive groups, or strata, and randomly sample from each of these groups. Using our mall example, our strata could be defined as male and female shoppers. Using stratified sampling, I can be sure that my final sample contains an equal number of male and female shoppers. This can be shown graphically in Figure 12.7.

Stat Facts _____

A **stratified sample** is obtained by dividing the population into mutually exclusive groups, or strata, and randomly sampling from each of these groups.

There are many different ways to establish strata from the population. Using the Academic Integrity survey, our strata could be defined as undergraduate and graduate students. If 20 percent of our college population are graduate students, I could use stratified sampling to ensure that 20 percent of my final sample are also graduate students. Other examples of criteria that can be used to divide the population into strata are age, income, or occupation.

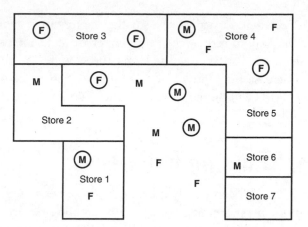

Figure 12.7

Stratified sampling.

Stratified sampling is helpful when it is important that the final sample has certain characteristics of the overall population. If we chose to use a simple random sample at the mall, the final sample may not have the desired proportion of males and females. This would lead to a biased sample if males feel differently about Deb's new look than females.

Sampling Errors

So far, we have stressed the benefits of drawing a sample from a population, rather than measure every member of the population. However, in statistics, as in life, there's no such thing as a free lunch. By relying on a sample, we expose ourselves to errors that can lead to inaccurate conclusions about the population.

The type of error that a statistician is most concerned about is called *sampling error,* which occurs when the sample measurement is different from the population measurement. Because the entire population is rarely measured in its entirety, the sampling error cannot be directly calculated. However, with inferential statistics, we'll be able to assign probabilities to certain amounts of sampling error later in Chapter 15.

Sampling errors occur because we might have the unfortunate luck of selecting a sample that is not a perfect match to the entire population. If the majority of mall shoppers really did like Deb's new look but we just happen to choose a bunch of morons for our sample who did not fully appreciate a good thing when they see it, Deb might never wear her new ponytail again.

Stat Facts

Sampling error occurs when the sample measurement is different from the population measurement. It is the result of selecting a sample that is not a perfect match to the entire population.

Sampling errors are expected and usually are a small price to pay to avoid measuring an entire population. One way to reduce the sampling error of a statistical study is to increase the size of the sample. In general, the larger the sample size, the smaller the sampling error. If you increase the sample size until it reaches the size of the population, then the sampling error will be reduced to 0. But in doing so, you forfeit the benefits of sampling.

Examples of Poor Sampling Techniques

The technique of sampling has been widely used, both properly and improperly, in the area of politics. One of the most famous mishaps with sampling occurred during the 1936 presidential race where the *Literary Digest* predicted Alf Landon to win the election over Franklin D. Roosevelt. Even if history is not your best subject, you would realize somebody had egg on his face after this election day. *Literary Digest* drew their sample from phonebooks and automobile registrations. The problem was that people with phones and cars in 1936 tended to be wealthier Republicans and were not representative of the entire voting population.

Another sampling blunder occurred in the 1948 presidential race when the Gallup poll predicted Thomas Dewey to be the winner over Harry Truman. The picture in Figure 12.8 shows a victorious Truman holding up the morning copy of the *Chicago Tribune* with the headline "Dewey Defeats Truman."

Figure 12.8

Dewey Defeats Truman.

The failure of the Gallup poll stemmed from the fact that there were a large number of undecided voters in the sample. It was wrongly assumed that these voters were representative of the decided voters who happened to favor Dewey. Truman easily won the election with 303 electoral votes compared to Dewey's 189.

CAUTION

Wrong Number

Have you ever participated in an online survey on a sports or news website that allowed you to view the results? These surveys can be fun and interesting, but the results needed to be taken with a grain of salt. That's because the respondents are self-selected, which means the sample is *not* randomly chosen. The results of these surveys are most likely biased, because the respondents would not be representative of the population at large. For example, people without Internet access would not be part of the sample and might respond differently than people with access to the Internet.

As you can see, choosing the proper sample is a critical step when using inferential statistics. Even a large sample size cannot hide the errors of choosing a sample that is not representative of the population at large. History has shown that large sample sizes are not needed to ensure accuracy. For example, the Gallup poll predicted that Richard Nixon would receive 43 percent of the votes for the 1968 presidential election when in fact he won 42.9 percent. This Gallup poll was based on a sample size of only 2,000, whereas the disastrous 1936 *Literary Digest* poll sampled 2,000,000 people. (Source: www.personal.psu.edu/faculty/g/e/gec7/Sampling.html)

Your Turn

1. A systematic sample is to be gathered from a local phone book with 75,000 names. If every k^{th} name in the phone book is to be selected, what value of k would be chosen to gather a sample size of 500?

2. Consider a population that is defined as every employee in a particular company. How could cluster sampling be used to gather a sample to participate in a survey involving employee satisfaction?

3. Consider a population that is defined as every employee in a particular company. How could stratified sampling be used to gather a sample to participate in a survey involving employee satisfaction?

The Least You Need to Know

◆ A simple random sample is a sample in which every member of the population has an equal chance of being chosen.

◆ In systematic sampling, every k^{th} member of the population is chosen for the sample, with value of k being approximately $\dfrac{N}{n}$.

- A cluster sample is a simple random sample of groups, or clusters, of the population. Each member of the chosen clusters would be part of the final sample.

- A stratified sample is obtained by dividing the population into mutually exclusive groups, or strata, and randomly sampling from each of these groups.

- Sampling error occurs when the sample measurement is different from the population measurement. It is the result of selecting a sample that is not a perfect match to the entire population.

Chapter

13

Sampling Distributions

In This Chapter

♦ Using sampling distributions of the mean and proportion

♦ Working with the central limit theorem

♦ Using the standard error of the mean and proportion

In Chapter 12, we praised the wonders of using samples in our statistical analysis because it was more efficient than measuring an entire population. In this chapter, we'll discover another benefit of using samples—sampling distributions.

Sampling distributions describe how sample averages behave. You may be surprised to hear they behave very well—even better than the populations from which they are drawn. Good behavior means we can do a pretty good job at predicting future values of sample means with a little bit of information. This might sound a little puzzling now, but by the end of this chapter you'll be shaking your head in utter amazement.

What Is a Sampling Distribution?

Let's say I want to perform a study to determine the number of miles the average person drives a car in one day. Because it's not possible to measure the driving patterns of every person in the population, I randomly choose a sample size of 10 ($n = 10$) qualified individuals and record how many miles they drove yesterday. I then chose another 10 drivers and record the same information. I do this three more times with the results in the following table.

Sample Number	Average Number of Miles (Sample Mean)
1	40.4
2	76.0
3	58.9
4	43.6
5	62.6

As you can see, each sample has its own mean value and each value is different. We can continue this experiment by selecting many more samples and observe the pattern of sample means. This pattern of sample means represents the sampling distribution for the number miles the average person drives in one day.

Sampling Distribution of the Mean

The distribution from the previous example represents the *sampling distribution of the mean* because the mean of each sample was the measurement of interest. This particular distribution has some interesting properties that will be discussed with the following example.

On a recent beach vacation, the resort we had chosen advertised a Ping Pong tournament, which caught the eye of my 15-year-old son, John, who has enough skill to normally beat his poor old father. After all, I only taught him how to play the game when he had to stand on top of a cooler to see over the table. As fate would have it, we were paired against each other and I found myself facing game point with the score tied. Swallowing my normally overzealous competitive spirit, I attacked the ball with a motion that somewhat resembles a person having an epileptic seizure and hit the ball right into the net. But that was a small price to pay since John's pride was saved as well as the rest of my vacation week. The things we do for our children.

Stat Facts

The **sampling distribution of the mean** refers to the pattern of sample means that will occur as samples are drawn from the population at large.

Anyway, using Ping-Pong balls to describe the way sample means behave, assume that I have 100 Ping-Pong balls in a container in which 20 balls are marked with the number 1, 20 marked with 2, 20 with 3, 20 with 4, and 20 with 5.

We can look at the probability distribution of this population in the following table.

Ball Number	Frequency	Relative Frequency	Probability
1	20	20/100	0.20
2	20	20/100	0.20
3	20	20/100	0.20
4	20	20/100	0.20
5	20	20/100	0.20

This is known as a *discrete uniform probability distribution* because each event has the same probability, as can be seen in Figure 13.1.

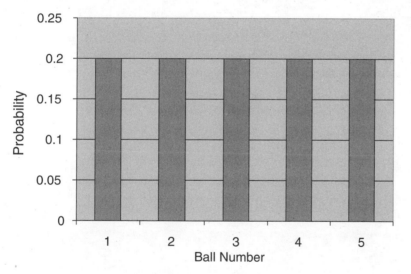

Figure 13.1

Discrete uniform probability distribution.

The mean and variance of a discrete uniform distribution can be calculated as follows:

$$\mu = \frac{1}{2}(a + b)$$

$$\sigma^2 = \frac{1}{12}(b - a)^2$$

where:

   a = minimum value of the distribution

   b = maximum value of the distribution

Stat Facts

A **discrete uniform probability distribution** is a distribution that assigns the same probability to each discrete event (and is discrete if it is countable).

For the Ping-Pong ball population:

$$\mu = \frac{1}{2}(1+5) = 3.0$$

$$\sigma^2 = \frac{1}{12}(5-1)^2 = \frac{16}{12} = 1.33$$

Keep these results in mind. We'll be referring back to them later in the chapter.

Now for my demonstration. With the balls evenly mixed, I select one ball, record the number, place it back in the container and then select a second ball, doing the same. This is my first sample with a size of 2 ($n = 2$). After doing this 25 times, I calculate the means of each sample and show the results in the following table.

Sampling Distribution of the Mean (n = 2)

Sample	First Ball	Second Ball	Sample Mean $\bar{x}$
1	1	3	2.0
2	2	1	1.5
3	2	1	1.5
4	1	1	1.0
5	4	2	3.0
6	1	3	2.0
7	1	2	1.5
8	3	1	2.0
9	2	5	3.5
10	1	3	2.0
11	3	3	3.0
12	4	2	3.0
13	5	2	3.5
14	3	1	2.0
15	1	4	2.5
16	4	4	4.0
17	2	2	2.0
18	2	2	2.0
19	1	1	1.0
20	2	5	3.5
21	1	2	1.5

Sample	First Ball	Second Ball	Sample Mean $\bar{x}$
22	5	5	5.0
23	3	2	2.5
24	5	5	5.0
25	2	1	1.5

I have a slight confession to make here. I really didn't buy 100 Ping-Pong balls and mark each one. The numbers from the previous table came from Excel's random number function that was discussed in Chapter 12.

We can convert this table into a relative frequency distribution, which is shown in the following table.

Wrong Number

Students often confuse sample size, n, and number of samples. In the previous example, the sample size equals 2 ($n = 2$), and the number of samples equals 25. In other words, we have 25 samples, each of size 2.

Sample Mean $\bar{x}$	Frequency	Relative Frequency	Probability
1.0	3	3/25	0.12
1.5	4	4/25	0.16
2.0	7	7/25	0.28
2.5	2	2/25	0.08
3.0	3	3/25	0.12
3.5	3	3/25	0.12
4.0	1	1/25	0.04
4.5	0	0/25	0.00
5.0	2	2/25	0.08

The previous table represents the sampling distribution of the mean for our Ping-Pong experiment with $n = 2$. We can show this distribution graphically in Figure 13.2.

I'm sure by now your highly inquisitive mind is screaming, "What happens to the sampling distribution if we increase the sample size?" That's an excellent question that will be addressed in the next section.

Figure 13.2

Sampling distribution of the mean for n = 2.

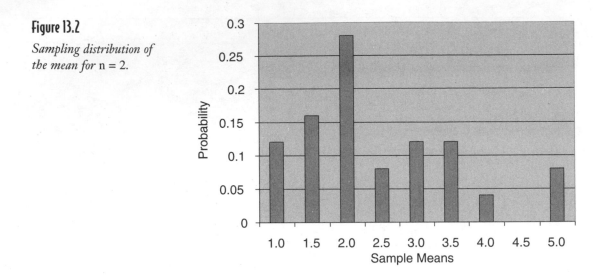

The Central Limit Theorem

As I mentioned earlier, sample means behave in a very special way. According to the *central limit theorem*, as the sample size, *n*, gets larger, the sample means tend to follow a normal probability distribution. This holds true regardless of the distribution of the population from which the sample was drawn. Amazing, you say.

Bob's Basics

The central limit theorem, in my humble opinion, is the most powerful concept for inferential statistics. It forms the foundation for many statistical models that are used today. It's a real good idea to cozy up to this theorem.

Stat Facts

According to the **central limit theorem,** as the sample size, *n*, gets larger, the sample means tend to follow a normal probability distribution and tend to cluster around the true population mean. This holds true regardless of the distribution of the population from which the sample was drawn.

As you look at Figure 13.2, you're probably scratching your head and thinking, "That distribution doesn't look like a normal curve, which I know is bell shaped and symmetrical." You're absolutely right because a sample size of 2 is generally not big enough for the central limit theorem to kick in.

Let's satisfy your curiosity and repeat my experiment by gathering 25 samples each consisting of 5 Ping-Pong balls (*n* = 5). I calculate the average of each sample and plot them in Figure 13.3.

Notice the impact increasing the sample size has on the shape of the sample distribution. It's starting to appear somewhat bell shaped with a little more symmetry. Let's look at sample sizes of 10 and 20 in Figures 13.4 and 13.5.

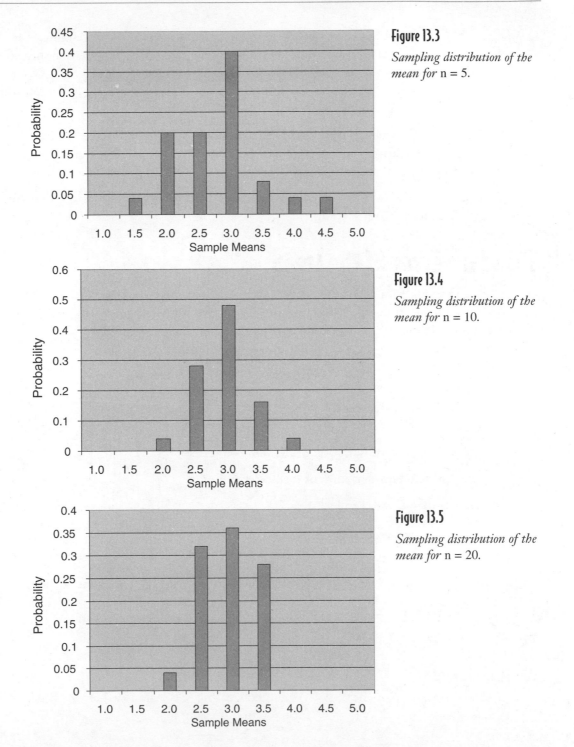

Figure 13.3

Sampling distribution of the mean for n = 5.

Figure 13.4

Sampling distribution of the mean for n = 10.

Figure 13.5

Sampling distribution of the mean for n = 20.

Note that as sample size increases, the sampling distribution tends to resemble a normal probability distribution. I don't know about you, but I find this pretty impressive considering the fact that the population that these samples were drawn from was not even close to being a normal distribution. If you recall, the Ping-Pong ball population followed a uniform distribution as shown in Figure 13.1.

Also, notice that as sample size increases, the sample means tend to cluster around the true population mean, which if you recall, we calculated as 3.0. This is another important feature of the central limit theorem.

Believe it or not, the central limit theorem has one more important feature that will be discussed in the next section.

Standard Error of the Mean

Notice in the last four figures, as the sample size increased, the distribution of sample means tended to converge closer together. In other words, as the sample size increased, the standard deviation of the sample means became smaller. According to the central limit theorem (here we go again!), the standard deviation of the sample means can be calculated as follows:

$$\sigma_{\bar{x}} = \frac{\sigma}{\sqrt{n}}$$

where:

$\sigma_{\bar{x}}$ = the standard deviation of the sample means

σ = the standard deviation of the population

n = sample size

The standard deviation of the sample means is formally known as the *standard error of the mean.*

Stat Facts

The **standard error of the mean** is the standard deviation of sample means. According to the central limit theorem, the standard error of the mean can be determined by $\sigma_{\bar{x}} = \frac{\sigma}{\sqrt{n}}$.

Recall that earlier in the chapter, in the section "Sampling Distribution of the Mean," we determined that the variance of the Ping-Pong ball population was 1.33. Therefore:

$$\sigma = \sqrt{\sigma^2} = \sqrt{1.33} = 1.15$$

We can now calculate the standard error of the mean for n = 2 in our example:

$$\sigma_{\bar{x}} = \frac{\sigma}{\sqrt{n}} = \frac{1.15}{\sqrt{2}} = 0.813$$

Bob's Basics

Students often confuse σ and $\sigma_{\bar{x}}$. The symbol σ, the standard deviation of the population, measures the variation within the population and was discussed back in Chapter 5. The symbol $\sigma_{\bar{x}}$, the standard error, measures the variation of the sample means and will decrease as the sample size increases.

The following table summarizes how the standard error varies with sample size in our Ping-Pong ball example.

Standard Error Varies with Sample Size

Sample Size	Standard Error
2	0.813
5	0.514
10	0.364
20	0.257

Why Does the Central Limit Theorem Work?

In this section, I explain why the central limit theorem behaves the way it does. If this concept does not interest you, feel free to skip over this section. I promise my feelings won't be hurt.

Going back to our original experiment with a sample size of 2, the following table shows all the 2-ball combinations that are possible along with the sample mean.

Sample	First Ball	Second Ball	Sample Mean $\bar{x}$
1	1	1	1.0
2	1	2	1.5
3	1	3	2.0
4	1	4	2.5
5	1	5	3.0
6	2	1	1.5
7	2	2	2.0
8	2	3	2.5

continues

continued

Sample	First Ball	Second Ball	Sample Mean $\bar{x}$
9	2	4	3.0
10	2	5	3.5
11	3	1	2.0
12	3	2	2.5
13	3	3	3.0
14	3	4	3.5
15	3	5	4.0
16	4	1	2.5
17	4	2	3.0
18	4	3	3.5
19	4	4	4.0
20	4	5	4.5
21	5	1	3.0
22	5	2	3.5
23	5	3	4.0
24	5	4	4.5
25	5	5	5.0

We can convert this table into a relative frequency distribution, which is shown in the following table.

Sample Mean $\bar{x}$	Frequency	Relative Frequency	Probability
1.0	1	1/25	0.04
1.5	2	2/25	0.08
2.0	3	3/25	0.12
2.5	4	4/25	0.16
3.0	5	5/25	0.20
3.5	4	4/25	0.16
4.0	3	3/25	0.12
4.5	2	2/25	0.08
5.0	1	1/25	0.04

The previous table represents the *theoretical sampling distribution of the mean* because it represents all the possible combinations of samples along with their respective probabilities. This distribution is shown graphically in Figure 13.6.

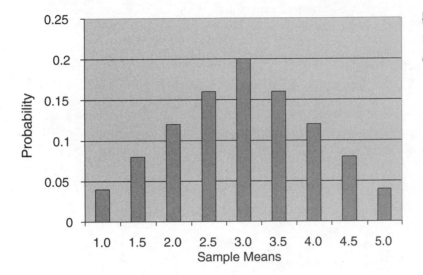

Figure 13.6

Theoretical sampling distribution of the mean.

You can see by this figure that the most common sample average is 3.0, whereas sample averages of 1.0 and 5.0 occur the least number of times. This is because there are more possible combinations of 2-ball samples that average to 3.0 (5 to be exact) than 2-ball samples that average to 1.0 or 5.0 (1 to be exact). In other words, we have 5 times the likelihood of drawing a 2-ball sample that averages 3.0 when compared to sample averages of 1.0 or 5.0.

Stat Facts

The **theoretical sampling distribution of the mean** displays all the possible sample means along with their classical probabilities. See Chapter 6 for a review of classical probability.

As we increase our sample size to 5, 10, and 20, the probability of drawing a sample with an average of 1.0 or 5.0 decreases while the probability of drawing a sample with an average of 3.0 increases. This explains why as sample size grows, more sample averages center around 3.0 and fewer around 1.0 and 5.0.

Putting the Central Limit Theorem to Work

I can just sense your need right now to do something really neat with the wonderful new tool. Look no further. If we know that the sample means follow the normal probability distribution and we also know the mean and standard deviation of that distribution, we can predict the likelihood that sample means will be greater or less than certain values.

For example, let's take our Ping-Pong ball experiment with n = 20. From the central limit theorem, we know that the sample means follow a normal distribution with:

$$\mu = 3.0$$

$$\sigma_{\bar{x}} = \frac{\sigma}{\sqrt{n}} = \frac{1.15}{\sqrt{20}} = 0.257$$

What is the probability that our next sample of 20 Ping-Pong balls will have a sample average of 3.4 or less? The sample mean distribution is shown in Figure 13.7 with the shaded region indicating the probability of interest.

Figure 13.7

Probability that next sample mean will be less than or equal to 3.4.

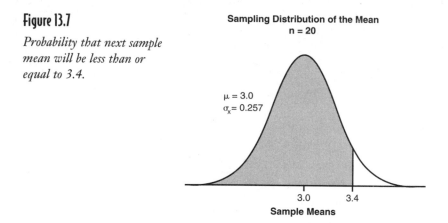

As we did in Chapter 11, we need to calculate the z-score. The equation looks slightly different because we are working with sample means, but in reality, it is identical to what we saw in Chapter 11.

$$z = \frac{\bar{x} - \mu}{\sigma_{\bar{x}}}$$

$$z_{3.4} = \frac{3.4 - 3.0}{0.257} = 1.56$$

Using the standard z-table in Appendix B:

$$P[\bar{x} \leq 3.4] = P[z \leq 1.56] = 0.9406$$

This probability is shown in Figure 13.8.

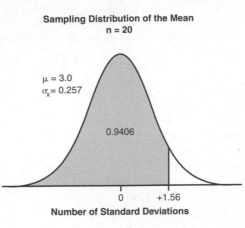

Sampling Distribution of the Mean
n = 20

μ = 3.0
σ$_x$ = 0.257

0.9406

0   +1.56

Number of Standard Deviations

Figure 13.8

Probability that next sample mean will be less than or equal to 1.56 standard deviations from the population mean.

According to the shaded region, the probability that our next sample of 20 Ping-Pong balls will have a sample mean of 3.4 or less is approximately 94 percent.

Using the Central Limit Theorem with an Unknown Population Mean

In our Ping-Pong ball experiment, we just happened to know that the population mean was 3.0. What do we do when the population mean is unknown? We can take the average of the sample means and use that as an approximation to the population mean.

To demonstrate, the following table shows the 25 sample means from our experiment when our sample size was 20 (*n* = 20).

Sample Means from 25 Ping-Pong Ball Samples

Sample	Sample Mean	Sample	Sample Mean
1	2.35	14	2.90
2	3.30	15	3.55
3	3.50	16	2.60
4	2.90	17	3.15
5	2.70	18	2.70
6	3.45	19	3.35
7	3.00	20	2.70
8	3.20	21	2.95
9	3.30	22	2.50

continues

continued

Sample	Sample Mean	Sample	Sample Mean
10	2.40	23	3.40
11	2.25	24	3.30
12	3.10	25	2.65
13	3.15		

If we add up the sample means and divide by 25, we obtain the grand average shown here:

$$\overline{\overline{x}} = \frac{\text{Sum of Sample Means}}{25} = \frac{74.35}{25} = 2.97$$

According to the central limit theorem, the population mean can be approximated by this grand average:

$$\mu \approx \overline{\overline{x}}$$

If we go back to our previous example to calculate the probability that the next sample mean will be less than or equal to 3.4, we get:

$$z_{3.4} = \frac{\overline{x} - \mu}{\sigma_{\overline{x}}} = \frac{3.4 - 2.97}{0.257} = 1.67$$

Using the standard z-table in Appendix B:

$$P[\overline{x} \le 3.4] = P[z \le 1.67] = 0.9527$$

This probability is slightly higher than the previous example because of the approximation of the population mean.

The power of the central limit theorem lies in the fact that you need little information about the distribution of the population to apply it. The sample means will behave very nicely as long as the sample size is large enough. It's a very versatile theorem that has countless applications in the real world. I knew you'd be impressed!

Sampling Distribution of the Proportion

The sample mean is not the only statistical measurement that is performed. What if I want to measure the percentage of teenagers in this country who would agree with the following statement: "My parents are an excellent resource when I'm looking for advice on an important matter in my life." Because each respondent has only two choices (agree or disagree), this experiment follows the binomial probability distribution, which was discussed in Chapter 9.

Calculating the Sample Proportion

My measurement of interest is the proportion of teenagers in my sample of size n, who will agree with the statement "My parents are an excellent resource when I'm looking for advice on an important matter in my life." The sample proportion, p_s, is calculated by:

$$p_s = \frac{\text{Number of Successes in the Sample}}{n}$$

Because I don't know the population proportion, p, who would agree with the statement, I need to collect data from samples and approximate the population proportion in the same fashion as I did earlier with sample means.

With proportion data, I want the sample size to be large enough so that I can use the normal probability distribution to approximate the binomial distribution. If you recall from Chapter 11, if $np \geq 5$ and $nq \geq 5$, we can use the normal distribution to approximate the binomial ($q = 1 - p$, the probability of a failure). I'm hopeful that p will be at least 5 percent (at least a few teenagers might listen to their parents), so if I choose $n = 150$, then:

$$np = (150)(0.05) = 7.5$$

$$nq = (150)(0.95) = 142.5$$

Suppose I choose 10 samples, each of size 150, and record the number of agreements (successes) in each sample in the table that follows.

> **CAUTION**
>
> **Wrong Number**
>
> It's important to remember that a proportion, either p or p_s, cannot be less than 0 or greater than 1. A common mistake that students make is when told that the proportion equals 10 percent, they set $p = 10$ rather than $p = 0.10$.

Sample	Number of Successes p_s	Sample Proportion
1	26	26/150 = 0.173
2	18	18/150 = 0.120
3	21	21/150 = 0.140
4	30	30/150 = 0.200
5	24	24/150 = 0.160
6	21	21/150 = 0.140
7	16	16/150 = 0.107
8	28	28/150 = 0.187
9	35	35/150 = 0.233
10	27	27/150 = 0.180

Next I average the sample proportions to approximate the population proportion, p:

$$p \approx \overline{p_s} = \frac{0.173 + 0.12 + 0.14 + 0.2 + 0.16 + 0.14 + 0.107 + 0.187 + 0.233 + 0.18}{10} = 0.164$$

Calculating the Standard Error of the Proportion

I now need to calculate the standard deviation of this sampling distribution, which is known as the *standard error of the proportion*, or σ_p, with the following equation:

$$\sigma_p = \sqrt{\frac{p(1-p)}{n}}$$

$$\sigma_p = \sqrt{\frac{0.164(1-0.164)}{150}} = \sqrt{0.000914} = 0.030$$

Stat Facts

The **standard error of the proportion** is the standard deviation of the sample proportions and can be calculated by

$$\sigma_p = \sqrt{\frac{p(1-p)}{n}}.$$

Now I'm ready to answer the age-old question, "What is the probability that from my next sample of 150 teenagers, 20 percent or less will agree with the statement: 'My parents are an excellent resource when I'm looking for advice on an important matter in my life'?" The shaded area in Figure 13.9 represents this probability, which displays the sampling distribution of the proportion for this example.

Figure 13.9

Sampling distribution of the proportion.

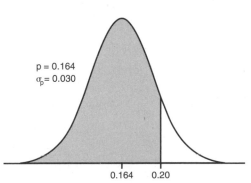

Sampling Distribution of the Proportion

p = 0.164
σ_p = 0.030

0.164   0.20

Sample Proportions

Because our sample size allows us the use the normal approximation to the binomial distribution, we now calculate the z-score for the proportion using the following equation:

$$z = \frac{p_s - p}{\sigma_p}$$

$$z_{0.20} = \frac{0.20 - 0.164}{0.030} = +1.20$$

Using the standard z-table in Appendix B:

$$P[p_s \leq 0.20] = P[z \leq 1.20] = 0.8849$$

This probability is also shown graphically in the shaded region in Figure 13.10.

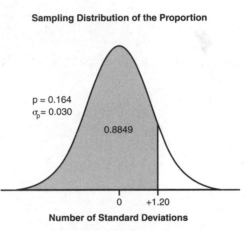

Sampling Distribution of the Proportion

p = 0.164
σ_p = 0.030

0.8849

0   +1.20

Number of Standard Deviations

Figure 13.10

Probability that next sample proportion will be less than or equal to 1.2 standard deviations from the population proportion.

According to our results, there is an 88.49 percent chance that 20 percent or fewer teenagers will agree with our statement from the next sample of size 150. Oh well, maybe when they get older they'll discover the real wisdom of their parents.

Your Turn

1. Calculate the standard error of the mean when …

   a. $\sigma = 10$, $n = 15$

   b. $\sigma = 4.7$, $n = 12$

   c. $\sigma = 7$, $n = 20$

2. A population has a mean value of 16.0 and a standard deviation of 7.5, calculate the following with a sample size of 9.

    a. $P[\bar{x} \leq 17]$

    b. $P[\bar{x} > 18]$

    c. $P[14.5 \leq \bar{x} \leq 16.5]$

3. Calculate the standard error of the proportion when …

    a. $p = 0.25$, $n = 200$

    b. $p = 0.42$, $n = 100$

    c. $p = 0.06$, $n = 175$

4. A population proportion has been estimated at 0.32. Calculate the following with a sample size of 160.

    a. $P[p_s \leq 0.30]$

    b. $P[p_s > 0.36]$

    c. $P[0.29 \leq p_s \leq 0.37]$

The Least You Need to Know

- The sampling distribution of the mean refers to the pattern of sample means that will occur as samples are drawn from the population at large.

- According to the central limit theorem, as the sample size, n, gets larger, the sample means tend to follow a normal probability distribution.

- According to the central limit theorem, as the sample size, n, gets larger, the sample means tend to cluster around the true population mean.

- The standard error of the mean is the standard deviation of sample means. According to the central limit theorem, the standard error of the mean can be determined by $\sigma_{\bar{x}} = \dfrac{\sigma}{\sqrt{n}}$.

- The standard error of the proportion is the standard deviation of the sample proportions and can be calculated by $\sigma_p = \sqrt{\dfrac{p(1-p)}{n}}$.

Chapter 14

Confidence Intervals

In This Chapter

- ◆ Interpreting the meaning of a confidence interval
- ◆ Calculating the confidence interval for the mean with large and small samples
- ◆ Introducing the Student's t-distribution
- ◆ Calculating the confidence interval for the proportion
- ◆ Determining sample sizes to attain a specific margin of error

So far we have learned how to collect a random sample and how sample means and sample proportions behave under certain conditions. Now we are ready to put those samples to work using confidence intervals.

One of the most important roles that statistics plays in today's world is to gather information from a sample and use that information to make a statement about the population from which it was chosen. We are using the sample as an estimate for the population. But just how good of an estimate is the sample providing us? The purpose of confidence intervals is to provide us with that answer.

Confidence Intervals for the Mean with Large Samples

In this section, we will learn how to construct a confidence interval for a population mean using a large sample size. By a large sample size, we are generally referring to $n \geq 30$. The first step in developing a confidence interval for a population involves the following discussion on estimators.

Estimators

The simplest estimate of a population is the *point estimate*, the most common being the sample mean. A point estimate is a single value that best describes the population of interest. This concept can be explained using the following example.

I think that my wife has been kidnapped and secretly replaced by a Deb look-alike who also happens to be completely addicted to the home shopping channel. No one or nothing in our household has escaped the products Deb has found on her new favorite TV show. She has purchased stuff for the car, the kitchen floor, the dog, her skin, her hair, and so on, and so on.

Stat Facts

A **point estimate** is a single value that best describes the population of interest, the sample mean being the most common. An **interval estimate** provides a range of values that best describes the population.

Suddenly "Diamonique Week" has become a major holiday in our household. I'm not really sure what Diamonique actually is but I suspect it is "available for a limited time only." Whenever I turn on any TV in the house, the channel always seems to be set to a very convincing home shopping channel-type person pleading with me to "Call now! Only 3 left!"

Anyway, let's say I want to estimate the average dollar value of an order for the home shopping channel population. If my sample average was $78.25, I could use that as my point estimate for the entire population of home shopping customers.

The advantage of a point estimate is that it is easy to calculate and easy to understand. The disadvantage, however, is that I have no clue as to how accurate this estimate really is.

To deal with this uncertainty, we can use an *interval estimate*, which provides a range of values that best describes the population. To develop an interval estimate, we need to learn about confidence levels in the following section.

Confidence Levels

A *confidence level* is the probability that the interval estimate will include the population parameter. A *parameter* is defined as a numerical description of a population characteristic, such as the mean.

Recall from Chapter 13, sample means will follow the normal probability distribution for large sample sizes. Let's say we want to construct an interval estimate with a 90 percent confidence level. This confidence level corresponds to a z-score from the standard normal table equal to 1.64 as shown in Figure 14.1.

Stat Facts

A **confidence interval** is a range of values used to estimate a population parameter and is associated with a specific confidence level. A **parameter** is data that describes a characteristic about a population.

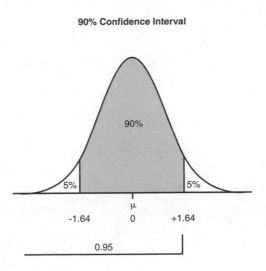

90% Confidence Interval

Figure 14.1

90 percent confidence level.

Bob's Basics

Notice that in Figure 14.1, 5 percent of the area under the curve lies to the right of +1.64 and 95 percent of the area under the curve lies to the left. That's why you see 0.9495 (close enough to 0.95) corresponding to a z-score of 1.64 in Table 3 of Appendix B. Remember, however, that $z = 1.64$ corresponds to a 90 percent confidence interval, the shaded region in the figure.

In general, we can construct a confidence interval around our sample mean using the following equations:

$\overline{x} + z_c\sigma_{\overline{x}}$ (upper limit of confidence interval)

$\overline{x} - z_c\sigma_{\overline{x}}$ (lower limit of confidence interval)

where:

$\overline{x}$ = the sample mean

z_c = the critical z-score, which is the number of standard deviations based on the confidence level

$\sigma_{\overline{x}}$ = the standard error of the mean (remember our friend from Chapter 13?)

The term $z_c\sigma_{\overline{x}}$ is referred to as the *margin of error*, or *E*, a phrase often referred to in polls and surveys.

Going back to our home shopping example, let's say from a sample of 32 customers, the average order is $78.25, and the population standard deviation is $37.50. (This represents the variation order to order in the population.) We can calculate our 90 percent confidence interval as follows:

$\overline{x} = \$78.25$

$n = 32$

$\sigma = \$37.50$

$z_c = 1.64$

$\sigma_{\overline{x}} = \dfrac{\sigma}{\sqrt{n}} = \dfrac{\$37.50}{\sqrt{32}} = \$6.63$

Upper limit =
$\overline{x} + 1.64\sigma_{\overline{x}} = \$78.25 + 1.64(\$6.63) = \$89.12$

Lower limit =
$\overline{x} - 1.64\sigma_{\overline{x}} = \$78.25 - 1.64(\$6.63) = \$67.38$

According to these results, our 90 percent confidence interval for this random sample of home shoppers is between $67.38 and $89.12 or ($67.38, $89.12). This interval is shown in Figure 14.2.

Figure 14.2

Interval estimate for the average dollar value of a home shopping order.

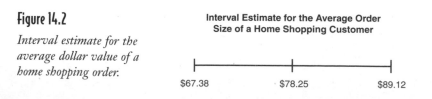

Interval Estimate for the Average Order
Size of a Home Shopping Customer

$67.38 $78.25 $89.12

Beware of the Interpretation of Confidence Interval!

As described in the previous section, a confidence interval is a range of values used to estimate a population parameter and is associated with a specific confidence level. A confidence interval needs to be described in the context of several samples. If we select 10 samples from our home shopping population and construct 90 percent confidence intervals around each of the sample means, then theoretically, 9 of the 10 intervals will contain the true population mean, which remains unknown. Figure 14.3 shows this concept.

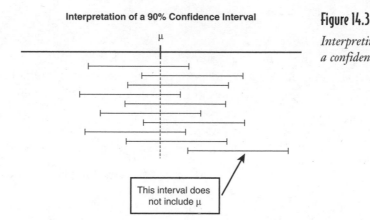

Interpretation of a 90% Confidence Interval

This interval does not include μ

Figure 14.3

Interpreting the definition of a confidence interval.

As you can see, Samples 1 through 9 have confidence intervals that include the true population mean, whereas Sample 10 does not.

CAUTION

Wrong Number _____

It is easy to misinterpret the definition of a confidence interval. For example, it is *not* correct to state that "there is a 90 percent probability that the true population mean is within the interval ($67.38, $89.12)." Rather, a correct statement would be that "there is a 90 percent probability that any given confidence interval from a random sample will contain the true population mean."

Because there is a 90 percent probability that any given confidence interval will contain the true population mean in the previous example, we have a 10 percent chance that it won't. This 10 percent value is known as the *level of significance*, α, which is represented by the total white area in both tails of Figure 14.4.

Figure 14.4

The level of significance.

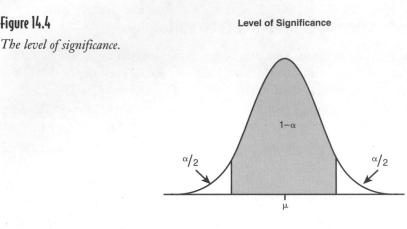

Level of Significance

The probability for the confidence interval is a complement to the significance level. For example, the significance level for a 95 percent confidence interval is 5 percent,

Stat Facts

The **level of significance** (α) is the probability of making a Type I error.

the significance level for a 99 percent confidence interval is 1 percent, and so on. In general, a $(1 - \alpha)$ confidence interval has a significance level equal to α.

We will revisit the level of significance in more detail in later chapters.

The Effect of Changing Confidence Levels

So far, we have only referred to a 90 percent confidence interval. However, we can choose other confidence levels to suit our needs. The following table shows our home shopping example with confidence levels of 90, 95, and 99 percent.

Confidence Intervals with Various Confidence Levels

Confidence Level	z_c	$\sigma_{\bar{x}}$	Sample Mean	Lower Limit	Upper Limit
90	1.64	$6.63	$78.25	$67.38	$89.12
95	1.96	$6.63	$78.25	$65.26	$91.24
99	2.57	$6.63	$78.25	$61.21	$95.29

From the previous table, you can se that there's a price to pay for increasing the confidence level—our interval estimate of the true population mean becomes wider and less precise. We have proven that once again, there is no free lunch with statistics. If we want more certainty that our confidence interval will contain the true population mean, that confidence interval will become wider.

Bob's Basics

I recommend that you confirm the z-scores in this table for yourself by checking with Table 3 in Appendix B. Practice makes perfect! Review Chapter 11 if you need to.

The Effect of Changing Sample Size

There is one way, however, to reduce the width of our confidence interval while maintaining the same confidence level. We can do this by increasing the sample size. There still is no free lunch, though, because increasing the sample size has a cost associated with it. Let's say we increase our sample size to include 64 home shoppers. This change will affect our standard error as follows:

$$\sigma_{\bar{x}} = \frac{\sigma}{\sqrt{n}} = \frac{\$37.50}{\sqrt{64}} = \$4.69$$

Our new 90 percent confidence interval for our original sample will be:

$$\bar{x} = \$78.25$$

$$n = 64$$

$$\sigma_{\bar{x}} = \$4.69$$

$$\text{Upper limit} = \bar{x} + 1.64\sigma_{\bar{x}} = \$78.25 + 1.64(\$4.69) = \$85.94$$

$$\text{Lower limit} = \bar{x} - 1.64\sigma_{\bar{x}} = \$78.25 - 1.64(\$4.69) = \$70.56$$

Increasing our sample size from 32 to 64 has reduced the 90 percent confidence interval from ($67.38, $89.12) to ($70.56, $85.94), which is a more precise interval.

Determining Sample Size for the Mean

We can also calculate a minimum sample size that would be needed to provide a specific margin of error. What sample size would we need for a 95 percent confidence interval that has a margin of error of $8.00 ($E = \$8.00$) in our home shopping example?

$$E = z\sigma_{\bar{x}}$$

$$E = \frac{z\sigma}{\sqrt{n}}$$

$$\sqrt{n} = \frac{z\sigma}{E}$$

$$n = \left(\frac{z\sigma}{E}\right)^2$$

$$n = \left(\frac{(1.96)(\$37.50)}{\$8.00}\right)^2 = 84.4 \approx 85$$

Therefore, to obtain a 95 percent confidence interval that ranges from \$78.25 – \$8.00 = \$70.25 to \$78.25 + \$8.00 = \$86.25 would require a sample size of 85 home shopping-addicted people.

Calculating a Confidence Interval When σ Is Unknown

Here's a simple section for you. (It's about time!) So far, all of our examples have assumed that we knew σ, the population standard deviation. What happens if σ is unknown? Don't panic, because as long as $n \geq 30$, we can substitute s, the sample standard deviation, for σ, the population standard deviation, and follow the same procedure as before. To demonstrate this technique, consider the following table that shows the order size in dollars of 30 home shoppers.

Home Shopping Sample (n = 30)

75	109	32	54	121	80	96	47	67	115
29	70	89	100	48	40	137	75	39	88
99	140	112	87	122	75	54	92	89	153

Using Excel, we can confirm that:

$$\bar{x} = \$84.47 \text{ and } s = \$32.98$$

A 99 percent confidence interval around this sample mean would be:

$$\bar{x} = \$84.47$$

$$n = 30$$

$$s = \$32.98$$

$$z_c = 2.57$$

$$\sigma^\mu_{\bar{x}} = \frac{s}{\sqrt{n}} = \frac{\$32.98}{\sqrt{30}} = \$6.02$$

We use $\sigma_{\bar{x}}^{\mu}$ to indicate that we have approximated the standard error of the mean by using s instead of σ. We statisticians just love to put little hats on top of letters.

$$\text{Upper limit} = \bar{x} + 2.57\sigma_{\bar{x}}^{\mu} = \$84.47 + 2.57(\$6.02) = \$99.94$$

$$\text{Lower limit} = \bar{x} - 2.57\sigma_{\bar{x}}^{\mu} = \$84.47 - 2.57(\$6.02) = \$69.00$$

See! That wasn't too bad.

Using Excel's CONFIDENCE Function

Excel has a pretty cool built-in function that calculates confidence intervals for us. The CONFIDENCE function has the following characteristics.

CONFIDENCE(alpha, standard_dev, size)

where:

alpha = the significance level of the confidence interval

standard_dev = the standard deviation of the population

size = sample size

For instance, Figure 14.5 shows the CONFIDENCE function being used to calculate the confidence interval for our original home shopping example.

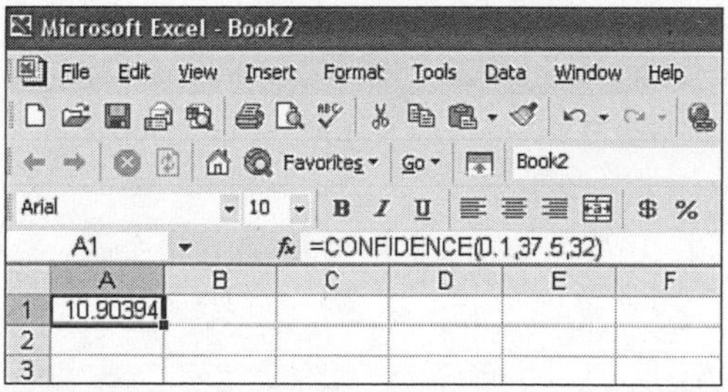

Figure 14.5

CONFIDENCE function in Excel for the home shopping sample.

Cell A1 contains the Excel formula =CONFIDENCE(0.1,37.5,32) with the result being 10.90394. This value represents the margin of error, or the amount to add and subtract from the sample mean, as follows:

$78.25 + $10.90 = $89.15

$78.25 − $10.90 = $67.35

This confidence interval is slightly different from the one calculated earlier in the chapter due to the rounding of numbers. This sure beats using tables and square root functions on the calculator.

Confidence Intervals for the Mean with Small Samples

So far, this entire chapter has dealt with the case where $n \geq 30$. I'm sure you are now wondering about how to construct a confidence interval when our sample size is less than 30. Well, as with many things in life, it depends.

With a small sample size, we lose the use of our faithful friend, the central limit theorem, and we need to assume that the population is normally (or approximately) distributed for all cases. The first case that we'll examine is when we know σ, the population standard deviation.

When σ Is Known

When σ is known, the procedure reverts back to the large sample size case. We can do this because we are now assuming the population is normally distributed. Let's construct a 95 percent confidence interval from the following home shopping sample of size 10.

Home Shopping Sample ($n = 10$)

75	109	32	54	121	80	96	47	67	115

We know the following information:

$\bar{x} = \$79.60$

$n = 10$

$\sigma = \$37.50$ (given from the original example)

$z_c = 1.96$

$$\sigma_{\bar{x}} = \frac{\sigma}{\sqrt{n}} = \frac{\$37.50}{\sqrt{10}} = \$11.86$$

Upper Limit = $\bar{x} + 1.96\sigma_{\bar{x}} = \$79.60 + 1.96(\$11.86) = \$102.85$

Lower Limit = $\bar{x} - 1.96\sigma_{\bar{x}} = \$79.60 - 1.96(\$11.86) = \$56.35$

Notice that the small sample size has resulted in a wide confidence interval. Again, we are assuming here that the population from which the sample was drawn is normally distributed, which is the first time we have made such an assumption in this chapter so far.

When σ Is Unknown

More often, we don't know the value of σ. Here, we make a similar adjustment that we made earlier and substitute *s*, the sample standard deviation, for σ, the population standard deviation. However, because of the small sample size, this substitution forces us to use a new probability distribution known as the Student's t-distribution (named in honor of you, the student).

The t-distribution is a continuous probability distribution with the following properties:

◆ It is bell shaped and symmetrical around the mean.

◆ The shape of the curve depends on the *degrees of freedom* (d.f.), which, when dealing with the sample mean, would be equal to *n* – 1.

◆ The area under the curve is equal to 1.0.

◆ The t-distribution is flatter than the normal distribution. As the number of degrees of freedom increase, the shape of the t-distribution becomes similar to the normal distribution as seen in Figure 14.6. With more than 30 degrees of freedom (a sample size of 30 or more), the 2 distributions are practically identical.

> **Random Thoughts**
>
> The Student's t-distribution was developed by William Gosset (1876–1937) while working for the Guinness Brewing Company in Ireland. He published his findings using the pseudonym Student. Now there's a rare statistical event—a bashful Irishman!

Stat Facts

The **degrees of freedom** are the number of values that are free to be varied given information, such as the sample mean, is known.

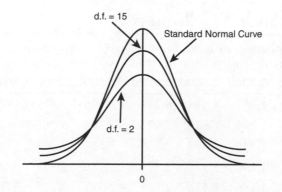

The t-Distribution Compared to the Normal Curve

d.f. = 15

Standard Normal Curve

d.f. = 2

0

Figure 14.6

The Student's t-distribution compared to the normal distribution.

Students often struggle with the concept of degrees of freedom, which represents the number of remaining free choices you have after something has been decided, such as the sample mean. For example, if I know that my sample of size 3 has a mean of 10, I can only vary 2 values ($n - 1$). After I set those 2 values, I have no control over the third value because my sample average must be 10. For this sample, I have 2 degrees of freedom.

We can now set up our confidence intervals for the mean using a small sample:

$\overline{x} + t_c \sigma_{\overline{x}}^{\mu}$ (upper limit of confidence interval)

$\overline{x} - t_c \sigma_{\overline{x}}^{\mu}$ (lower limit of confidence interval)

where:

t_c = critical t-value (can be found in Table 4 in Appendix B)

$\sigma_{\overline{x}}^{\mu} = \dfrac{s}{\sqrt{n}}$, the estimated standard error of the mean

To demonstrate this procedure, let's assume the population of home shopping orders follows a normal distribution and the following sample of size 10 was collected.

Home Shopping Sample from a Normal Distribution ($n = 10$)

| 29 | 70 | 89 | 100 | 48 | 40 | 137 | 75 | 39 | 88 |

With σ unknown, we will construct a 95 percent confidence interval around the sample mean.

To determine the value of t_c for this example, I need to calculate the number of degrees of freedom. Because $n = 10$, I have $n - 1 = 9$ d.f. This corresponds to $t_c = 2.262$, which is underlined in the following table taken from Table 4 in Appendix B.

Excerpt from the Student's T-Distribution

Selected right tail areas with confidence levels underneath

conf lev df	0.2000 0.6000	0.1500 0.7000	0.1000 0.8000	0.0500 0.9000	0.0250 0.9500	0.0100 0.9800	0.0050 0.9900	0.0010 0.9980	0.0005 0.9990
1	1.376	1.963	3.078	6.314	12.706	31.821	63.657	318.31	636.62
2	1.061	1.386	1.886	2.920	4.303	6.965	9.925	22.327	31.599
3	0.978	1.250	1.638	2.353	3.182	4.541	5.841	10.215	12.924
4	0.941	1.190	1.533	2.132	2.776	3.747	4.604	7.173	8.610
5	0.920	1.156	1.476	2.015	2.571	3.365	4.032	5.893	6.869

Selected right tail areas with confidence levels underneath								
conf lev	0.2000	0.1500	0.1000	0.0500	0.0250	0.0100	0.0050	0.0010	0.0005
df	0.6000	0.7000	0.8000	0.9000	0.9500	0.9800	0.9900	0.9980	0.9990
6	0.906	1.134	1.440	1.943	2.447	3.143	3.707	5.208	5.959
7	0.896	1.119	1.415	1.895	2.365	2.998	3.499	4.785	5.408
8	0.889	1.108	1.397	1.860	2.306	2.896	3.355	4.501	5.041
9	0.883	1.100	1.383	1.833	2.262	2.821	3.250	4.297	4.781
10	0.879	1.093	1.372	1.812	2.228	2.764	3.169	4.144	4.587
11	0.876	1.088	1.363	1.796	2.201	2.718	3.106	4.025	4.437
12	0.873	1.083	1.356	1.782	2.179	2.681	3.055	3.930	4.318

We next need to calculate the sample mean and sample standard deviation, which, according to Excel, are as follows:

$$\bar{x} = \$71.50 \text{ and } s = \$33.50$$

We can now approximate the standard error of the mean:

$$\sigma^{\mu}_{\bar{x}} = \frac{s}{\sqrt{n}} = \frac{\$33.50}{\sqrt{10}} = \$10.59$$

and can construct our 95 percent confidence interval:

$$\text{Upper limit} = \bar{x} + t_c \, \sigma^{\mu}_{\bar{x}} = \$71.50 + 2.262(\$10.59) = \$95.45$$
$$\text{Lower limit} = \bar{x} - t_c \, \sigma^{\mu}_{\bar{x}} = \$71.50 - 2.262(\$10.59) = \$47.55$$

Now that wasn't too bad!

Stat Facts

The t-distribution can be used when all of the following conditions have been met:

- The population follows the normal (or approximately normal) distribution.
- The sample size is less than 30.
- The population standard deviation, σ, is unknown and must be approximated by s, the sample standard deviation.

The following table summarizes this chapter to this point to help you decide which method to use to construct a confidence interval around the mean.

Summary of Confidence Intervals of the Mean

Conditions	Confidence Interval
$n \geq 30$, σ is known Any Population	$\bar{x} \pm z_c \sigma_{\bar{x}}$
$n \geq 30$, σ is unknown Any Population	$\bar{x} \pm z_c \sigma^{\mu}_{\bar{x}}$
$n < 30$, σ is known Normal Population	$\bar{x} \pm z_c \sigma_{\bar{x}}$
$n < 30$, σ is unknown Normal Population	$\bar{x} \pm t_c \sigma^{\mu}_{\bar{x}}$

That ends our discussion on confidence intervals around the mean. Next on the menu are proportions!

Confidence Intervals for the Proportion with Large Samples

We can also estimate the proportion of a population by constructing a confidence interval from a sample. Recall from Chapter 13, proportion data follow the binomial distribution that can be approximated by the normal distribution under the following conditions:

$$np \geq 5 \text{ and } nq \geq 5$$

where:

p = the probability of a success in the population

q = the probability of a failure in the population ($q = 1 - p$)

Suppose I want to estimate the proportion of home shopping customers who are female based on the results of a sample. Recall from the previous chapter that we can calculate the proportion of a sample using:

$$p_s = \frac{\text{Number of Successes in the Sample}}{n}$$

Calculating the Confidence Interval for the Proportion

The confidence interval around the sample proportion can be calculated by:

$P_s + z_c \sigma_p$ (upper limit of confidence interval)

$P_s - z_c \sigma_p$ (lower limit of confidence interval)

where σ_p is the standard error of the proportion (which is the standard deviation of the sample proportions) using:

$$\sigma_p = \sqrt{\frac{p(1-p)}{n}}$$

There's extra credit for anyone who can see a problem arising here. Our challenge is that we are trying to estimate p, the population proportion, but we need a value for p to set up the confidence interval. Our solution—estimate the standard error by using the sample proportion as an approximation for the population proportion as follows:

$$\sigma_{\bar{x}}^{\mu} = \sqrt{\frac{p_s(1-p_s)}{n}}$$

We now can construct a confidence interval around the sample proportion by:

$P_s + z_c \sigma_{\bar{x}}^{\mu}$ (upper limit of confidence interval)

$P_s - z_c \sigma_{\bar{x}}^{\mu}$ (lower limit of confidence interval)

Let's put these equations to work. In my efforts to estimate the proportion of female home shopping customers, I sample 175 random customers, of which 110 are female. I can now calculate p_s, the sample proportion:

$$p_s = \frac{\text{Number of Successes in the Sample}}{n} = \frac{110}{175} = 0.629$$

The estimated standard error of the proportion would be:

$$\sigma_{\bar{x}}^{\mu} = \sqrt{\frac{p_s(1-p_s)}{n}} = \sqrt{\frac{(0.629)(0.371)}{175}} = 0.0365$$

We are now ready to construct a 90 percent confidence interval around our sample proportion ($z_c = 1.64$):

Upper limit = $p_s + 1.64\sigma_p = 0.629 + 1.64(0.0365) = 0.689$

Lower limit = $p_s - 1.64\sigma_p = 0.629 - 1.64(0.0365) = 0.569$

Our 90 percent confidence interval for the proportion of female home shopping customers is (0.569, 0.689). Deb must be in there somewhere!

Determining Sample Size for the Proportion

Almost done. Just as we did for the mean, we can determine a required sample size that would be needed to provide a specific margin of error. What sample size would we need for a 99 percent confidence interval that has a margin of error of 6 percent ($E = 0.06$) in our home shopping example? The formula to calculate n, the sample size is:

$$n = pq\left(\frac{z_c}{E}\right)^2$$

Notice that we need a value for p and q. If we don't have a preliminary estimate of the values, set $p = q = 0.50$. Because half the population is female, that sounds like a good strategy to me.

$$n = (0.50)(0.50)\left(\frac{2.57}{0.06}\right)^2 = 459$$

Therefore, to obtain a 99 percent confidence interval that provides a margin of error no more than 6 percent would require a sample size of 459 home shoppers.

> ### Random Thoughts
>
> The reason we use $p = q = 0.50$ if we don't have an estimate of the population proportion is that these values provide the largest sample size when compared to other combinations of p and q. It's like being penalized for not having specific information about your population. This way you are sure your sample size is large enough, regardless of the population proportion.

Your Turn

1. Construct a 97 percent confidence interval around a sample mean of 31.3 taken from a population that is not normally distributed with a standard deviation of 7.6 using a sample of size 40.

2. What sample size would be necessary to ensure a margin of error of 5 for a 98 percent confidence interval taken from a population that is not normal, which has a population standard deviation of 15?

3. Construct a 90 percent confidence interval around a sample mean of 16.3 taken from a population that is not normally distributed with a population standard deviation of 1.8 using a sample of size 10.

4. The following sample of size 30 was taken from a population that is not normally distributed:

10	4	9	12	5	17	20	9	4	15
11	12	16	22	10	25	21	14	9	8
14	16	20	18	8	10	28	19	16	15

   Construct a 90 percent confidence interval around the mean.

5. The following sample of size 12 was taken from a population that is normally distributed and that has a population standard deviation of 12.7:

   37 48 30 55 50 46 40 62 50 43 36 66

   Construct a 94 percent confidence interval around the mean.

6. The following sample of size 11 was taken from a population that is normally distributed:

   121 136 102 115 126 106 115 132 125 108 130

   Construct a 98 percent confidence interval around the mean.

7. The following sample of size 11 was taken from a population that is not normally distributed:

   87 59 77 65 98 90 84 56 75 96 66

   Construct a 99 percent confidence interval around the mean.

8. A sample of 200 light bulbs was tested and it was found that 11 were defective. Calculate a 95 percent confidence interval around this sample proportion.

9. What sample size would be needed to construct a 96 percent confidence interval around the proportion for voter turnout during the next election that would provide a margin of error of 4 percent? Assume the population proportion has been estimated at 55 percent.

The Least You Need to Know

◆ A confidence interval is a range of values used to estimate a population parameter and is associated with a specific confidence level.

◆ A confidence level is the probability that the interval estimate will include the population parameter, such as the mean.

◆ Increasing the confidence level results in the confidence interval to become wider and less precise.

◆ Increasing the sample size reduces the width of the confidence interval, which increases precision.

◆ Use the t-distribution to construct a confidence interval when the population follows the normal (or approximately normal) distribution, the sample size is less than 30, and the population standard deviation, σ, is unknown.

◆ Use the normal distribution to construct a confidence interval around the sample proportion when $np \geq 5$ and $nq \geq 5$.

Introduction to Hypothesis Testing

In This Chapter

- ◆ Formulating the null and alternative hypothesis
- ◆ Distinguishing between a one-tail and two-tail hypothesis test
- ◆ Controlling the probability of a Type I and Type II error
- ◆ Determining the boundaries for the rejection region for the hypothesis test
- ◆ Stating the conclusion of the hypothesis test

Now that we know how to make an estimate of a population parameter, such as a mean, using a sample and a confidence interval, we are ready to move on to the heart and soul of inferential statistics: hypothesis testing.

One thing that statisticians like to do is to make a statement about a population parameter, collect a sample from that population, measure the sample and declare, in a scholarly manner, whether or not the original statement has been supported by the sample. This, in a nutshell, is what hypothesis testing is all about. Of course, I've included a few juicy details. Without them, this would be one short chapter!

The purpose of this particular chapter is to just introduce the basic concept of hypothesis testing. The following two chapters will then get into more specific examples of how we put hypothesis testing to work. Stay tuned!

Hypothesis Testing—the Basics

In the statistical world, a *hypothesis* is an assumption about a population parameter. Examples of hypotheses (that's plural for hypothesis) include the following:

- The average adult drinks 1.7 cups of coffee per day.

- Twelve percent of undergraduate students will go directly to graduate school after graduation.

- No more than 2 percent of our products that we sell to customers are defective.

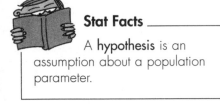

Stat Facts

A **hypothesis** is an assumption about a population parameter.

In each case, we have made a statement about the population that may or may not be true. The purpose of hypothesis testing is to make a statistical conclusion about accepting or not accepting such statements. To further explain this concept, I present the following story.

I am man enough to admit that I am deathly afraid of snakes. That's why I did not hesitate to express my panic when Sam, Deb's oldest teenage son, brought home a snake that he caught (and Deb wholeheartedly agreed to let him keep in his bedroom).

Well, my worst nightmare came true the following morning. The snake had pushed off the top of the cage overnight and was loose somewhere in the house. I guess Sam never heard the story of the mommy snake that once lifted a Volkswagen Beetle off of her baby snake to save it.

I won't name names here but *somebody's wife* suggested that we put a mouse in Sam's room to attract the snake so we could catch it. I thought this was a very good joke until a white mouse showed up in Sam's room later that day posing as "snake bait."

That night, I lay in bed under high alert (i.e., at least one eye always open and ears finely tuned for a hissing noise) while Deb lay calmly snoring next to me.

The next morning, I discovered that I had a *new* worst nightmare. The mouse had chewed its way out of its container overnight and it, too, was loose somewhere in the house. I now had *two* wild animals roaming freely in the places where I eat, sleep, and watch TV. By this time I'm frantically looking through the phonebook for a motel that specifically prohibits all snakes and mice. Deb thought I was "overreacting."

That night I lay in my bed in the fetal position to protect my vital organs and keep my arms and legs away from the side of the bed while Deb lay calmly snoring next to me.

Anyway, let's try to tie this sci-fi tale to hypothesis testing. Let's say that my hypothesis is that it will take an average of 6 days to capture a loose snake in a house. In other words, I would like to test my belief that the population mean, μ, is equal to 6 days. I do so by gathering a sample of people who have had a loose snake in their home and calculate the average number of days required to capture it. Suppose the sample average is 6.1 days. The hypothesis test will then tell me whether or not 6.1 days is significantly different from 6.0 days or if the difference is merely due to chance. More details to follow!

The Null and Alternative Hypothesis

Every hypothesis test has both a null hypothesis and an alternative hypothesis. The *null hypothesis*, denoted by H_0, represents the status quo and involves stating the belief that the mean of the population is $\geq$, =, or $\leq$ a specific value. The null hypothesis is believed to be true unless there is overwhelming evidence to the contrary. In this example, my null hypothesis would be stated as:

$$H_0 : \mu = 6.0$$

The *alternative hypothesis*, denoted by H_1, represents the opposite of the null hypothesis and holds true if the null hypothesis is found to be false. The alternative hypothesis always states the mean of the population is <, $\neq$, or > a specific value. In this example, my alternative hypothesis would be stated as:

$$H_1 : \mu \neq 6.0$$

The following table shows the three valid combinations of the null and alterative hypothesis.

Null Hypothesis	Alternative Hypothesis
$H_0 : \mu = 6.0$	$H_1 : \mu \neq 6.0$
$H_0 : \mu \geq 6.0$	$H_1 : \mu < 6.0$
$H_0 : \mu \leq 6.0$	$H_1 : \mu > 6.0$

> ### Random Thoughts
>
> Some textbooks will use the convention that the null hypothesis will always be stated as =
> and will never use ≤ or ≥. Choosing either method of stating your hypothesis will not affect
> the statistical analysis. Just be consistent with the convention that you decide to use.

Note that the alternative hypothesis never is associated with ≤, =, or ≥. Selecting the proper combination is the topic of the next section.

Stating the Null and Alternative Hypothesis

Care needs to be taken on how to state the null and alternative hypothesis. Your choice will depend on the nature of the test and the motivation of the person conducting it.

If the purpose is to test that the population mean is equal to a specific value, such as our snake example, assign this statement as the null hypothesis, which results in the following:

$H_0: \mu = 6.0$

$H_1: \mu \neq 6.0$

Often hypothesis testing is performed by researchers who want to prove that their discovery is an improvement over current products or procedures. For example, if I invented a golf ball that I claimed would increase your distance off the tee by more than 20 yards, I would set up my hypothesis as follows:

$H_0: \mu \leq 20$

$H_1: \mu > 20$

Note that I used the alternative hypothesis to represent the claim that I want to prove statistically so that I can make a fortune selling these balls to desperate golfers such as myself. Because of this, the alternative hypothesis is also known as the research hypothesis, because it represents the position that the researcher wants to establish.

Two-Tail Hypothesis Test

A *two-tail hypothesis test* is used whenever the alternative hypothesis is expressed as ≠. Our snake example would involve a two-tail test because the alternative hypothesis is stated as $H_1: \mu \neq 6.0$. This test can be shown graphically in Figure 15.1 which, as you can see, is considered a two-tail hypothesis test.

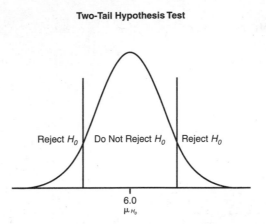

Two-Tail Hypothesis Test

Reject H_0 Do Not Reject H_0 Reject H_0

6.0
μ_{H_0}

Mean No. of Days to Catch a Snake

Figure 15.1

Two-tail hypothesis test.

The curve in the figure represents the sampling distribution of the mean for the number of days to catch a snake. The mean of the population, assumed to be 6.0 days according to the null hypothesis, is the mean of the sampling distribution and is designated by μ_{H_0}.

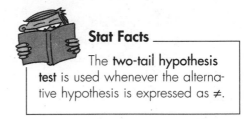

Stat Facts

The **two-tail hypothesis test** is used whenever the alternative hypothesis is expressed as ≠.

The procedure is as follows:

◆ Collect a sample of size *n*, and calculate the test statistic, which in this case is the sample mean.

◆ Plot the sample mean on the x-axis of the sampling distribution curve.

◆ If the sample mean falls within the white region, we do not reject H_0. That is, we do not have enough evidence to support H_1, the alternative hypothesis, which states that the population mean is not equal to 6.0 days.

◆ If the sample mean falls in either shaded region, otherwise known as the *rejection region*, we reject H_0. That is, we have enough evidence to support H_1, which results in our belief that the true population mean is not equal to 6.0 days.

Because there are two rejection regions in this figure, we have a two-tail hypothesis test. We will discuss how to determine the boundaries for the rejection regions shortly.

Wrong Number _____

The only two statements that we can make about the null hypothesis are that we …

◆ Reject the null hypothesis.

◆ Do not reject the null hypothesis.

Because our conclusions are based on a sample, we will never have enough evidence to accept the null hypothesis. It's a much safer statement to say that we do not have enough evidence to reject H_0. We can use the analogy of the legal system to explain. If a jury finds a defendant "not guilty," they are not saying the defendant is innocent. Rather, they are saying that there is not enough evidence to prove guilt.

One-Tail Hypothesis Test

A _one-tail hypothesis test_ involves the alternative hypothesis being stated as < or >. My golf ball example results in a one-tail test because the alternative hypothesis being expressed as $H_1: \mu > 20$ and is shown in Figure 15.2.

Figure 15.2

One-tail hypothesis test.

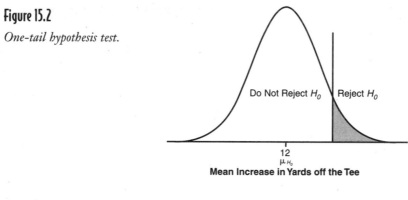

Do Not Reject H_0 | Reject H_0

12
μ_{H_0}
Mean Increase in Yards off the Tee

Stat Facts _____

The **one-tail hypothesis test** is used when the alternative hypothesis is being stated as < or >.

Here, there is only one rejection region, which is the shaded area on the right tail of the distribution. We follow the same procedure outlined for the two-tail test and plot the sample mean, which represents the average increase in distance from the tee with my new golf ball. Two possible scenarios exist.

♦ If the sample mean falls within the white region, we do not reject H_0. That is, we do not have enough evidence to support H_1, the alternative hypothesis, which states that my golf ball increased distance off the tee by more than 20 yards. There goes my fortune down the drain!

♦ If the sample mean falls in the rejection region, we reject H_0. That is, we have enough evidence to support H_1, which confirms my claim that my new golf ball will increase distance off the tee by more than 20 yards. Early retirement, here I come!

Bob's Basics _____

For a one-tail hypothesis test, the rejection region will always be consistent with the direction of the inequality for H_1. For $H_1 : \mu > 20$, the rejection region will be in the right tail of the sampling distribution. For $H_1 : \mu < 20$, the rejection region will be in the left tail.

Now that we have covered the basics of hypothesis testing, we are now ready to consider errors that can occur due to sampling.

Type I and Type II Errors

Remember that the purpose of the hypothesis test is to verify the validity of a claim about a population based on a single sample. Because we are relying on a sample, we expose ourselves to the risk that our conclusions about the population will be wrong.

Using the golf ball example, suppose that my sample falls within the "Reject H_0" region of the last figure. That is, according to the sample, my golf ball increases distance off the tee by more than 20 yards. But what if the true population mean is actually much less than 20 yards? This can occur primarily because of sampling error, which was discussed in Chapter 12. This type of error, when we reject H_0 when, in reality, it's true, is known as a *Type I error*. The probability of making a Type I error is known as α, the level of significance, which was first introduced in Chapter 14.

We also can experience another type of error with hypothesis testing. Let's say the golf ball sample fell within the "Do Not Reject H_0" region of the last figure. That is, according to the sample, my golf ball does not increase the distance off the tee by more than 20 yards. But what if the true population mean is actually much more than 20 yards? This type of error, when we do not reject H_0 when, in reality, it's false, is known as a *Type II error*. The probability of making a Type II error is known as β, the *power* of the hypothesis test.

The following table summarizes the two types of hypothesis errors.

	H_0 **Is True**	H_0 **Is False**
Reject H_0	Type I Error	Correct Outcome
	P[Type I Error] = α	
Do Not Reject H_0	Correct Outcome	Type II Error
		P[Type II Error] = β

Stat Facts

A **Type I error** occurs when the null hypothesis is not accepted, when in reality, it is true. A **Type II error** occurs when the null hypothesis is accepted when, in reality, it is not true.

Normally, with hypothesis testing, we decide on a value for α that is somewhere between 0.01 and 0.10 before we collect the sample. The value of β can then be calculated, but that topic goes beyond the scope of this book. Be grateful for this because that concept is very complicated!

Let's put these concepts to work now and do some real hypothesis testing!

Random Thoughts

Ideally, we would like the values of α and β to be as small as possible. However, for a given sample size, reducing the value of α will result in an increase in the value of β. The opposite also holds true. The only way to reduce both α and β simultaneously is to increase the sample size. Once the sample size has been increased to the size of the population, the values of α and β will be 0. However, as we discussed in Chapter 12, this is not a recommended strategy.

Example of a Two-Tail Hypothesis Test

The hypotheses for the snake example was stated as:

$H_0 : \mu = 6.0$

$H_1 : \mu \neq 6.0$

where μ = the mean number of days to catch a loose snake in a home.

Lets' say that I know that the standard deviation of the population, σ, is 0.5 days, and my sample size to test the hypothesis, n, is 30 homes. (Please don't ask me how I'm going to find 30 homes with loose snakes. I'm making this up as I go along, so just humor me.) We'll also set $\alpha = 0.05$, which means I'm willing to accept a 5 percent chance of committing a Type I error. Our first step is to calculate the standard error of the mean, $\sigma_{\bar{x}}$. If you remember from Chapter 13, the equation is:

$$\sigma_{\bar{x}} = \frac{\sigma}{\sqrt{n}} = \frac{0.50}{\sqrt{30}} = 0.0913 \text{ days}$$

Let's assume the sample mean from the 30 homes is 6.1 days. What is our conclusion about our estimate of the population mean, μ?

To answer this, we next have to determine the critical z-score, which corresponds to $\alpha = 0.05$. Because this is a two-tail test, this area needs to be evenly divided between both tails with each tail receiving $\alpha/2 = 0.025$. According to Figure 15.3, we need to find the critical z-score that corresponds to the area $0.950 + 0.025 = 0.975$. As you can see, the 0.950 area is derived from $1 - \alpha$.

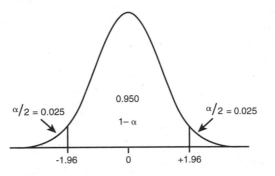

Figure 15.3

Critical z-score for $\alpha = 0.05$.

Using Table 3 in Appendix B, we look for the closest value to 0.9750 in the body of the table. We can find this value looking across column 1.9 and down row 0.06 to arrive at the z-score of +1.96 for the right tail and −1.96 for the left tail.

Using the Scale of the Original Variable

This section will determine the rejection region using the scale of the original variable, which in this case is the number of days. To calculate the upper and lower limits of the rejection region, we use the following equations. Recall from Chapter 14 that we use the z-scores from the standard normal distribution when $n \geq 30$ and σ is known.

Limits of rejection region = $\mu_{H_0} + z_c \sigma_{\bar{x}}$

where $\mu_{H_0} =$ population mean assumed by the null hypothesis.

For our snake example:

$$\text{Upper limit} = \mu_{H_0} + z_c \sigma_{\bar{x}} = 6.0 + (1.96)(0.0913) = 6.18 \text{ days}$$

$$\text{Lower limit} = \mu_{H_0} + z_c \sigma_{\bar{x}} = 6.0 + (-1.96)(0.0913) = 5.82 \text{ days}$$

Because our sample mean is 6.1 days, this falls within the "Do Not Reject H_0" region as shown in Figure 15.4. Our conclusion is that the difference between 6.1 days and 6.0 days is merely due to chance variation and we have support that the population mean is 6 days.

Figure 15.4

Hypothesis test for the snake example (original variable scale).

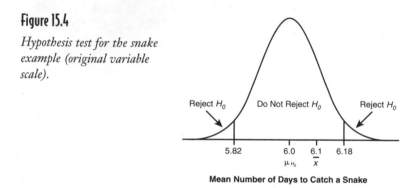

Mean Number of Days to Catch a Snake

Using the Standardized Normal Scale

We can arrive at the same conclusion by setting up the boundaries for the rejection region using the standardized normal scale. We do this by calculating the z-score that corresponds to the sample mean as follows:

$$z = \frac{\bar{x} - \mu_{H_0}}{\sigma_{\bar{x}}} = \frac{6.1 - 6.0}{0.0913} = +1.09$$

Bob's Basics

Be sure to distinguish between the calculated z-score and the critical z-score. The calculated z-score, z, represents the number of standard deviations between the sample mean and μ_{H_0}, the population mean according to the null hypothesis. The critical z-score, z_c, is based on the significance level, α, and determines the boundary for the rejection region.

Figure 15.5 shows this result graphically. Because the calculated z-score of +1.09 is within the "Do Not Reject H_0" region, the conclusions on the 2 techniques are consistent.

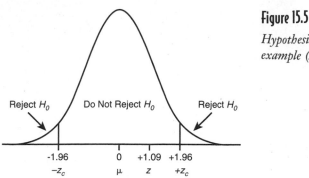

Example of a One-Tail Hypothesis Test

Because the alternative hypothesis for the golf ball example was formulated as > 20, this becomes a one-tail test. The hypothesis for this example was stated as:

$H_0 : \mu \le 20$

$H_1 : \mu > 20$

where μ = the mean increase in yards off the tee using my new golf ball.

Let's say that I know that the standard deviation of the population, σ, is 5.3 yards and my sample size to test the hypothesis, n, is 40 golfers. For this example, we'll set $\alpha = 0.01$. The standard error of the mean, $\sigma_{\bar{x}}$, will now be equal to:

$$\sigma_{\bar{x}} = \frac{\sigma}{\sqrt{n}} = \frac{5.3}{\sqrt{40}} = 0.838 \text{ yards}$$

Let's assume the sample mean from the 40 golfers is 22.5 yards. What is our conclusion about our estimate of the population mean, μ?

Once again, we next have to determine the critical z-score, which corresponds to $\alpha = 0.01$. Because this is a one-tail test, this entire area needs to be in one rejection region on the right side of the distribution. According to Figure 15.6, we need to find the z-score that corresponds to the area 0.99 or $1 - \alpha$.

Using Table 3 in Appendix B, we look for the closest value to 0.9900 in the body of the table, which results in a critical z-score of 2.33.

Figure 15.6

Critical z-score for $\alpha = 0.01$.

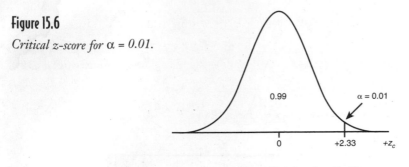

Number of Standard Deviations from the Mean

To calculate the limit for this rejection region using the scale of the original variable, we use:

$$\text{Limit} = \mu_{H_0} + z_c \sigma_{\bar{x}} = 20 + (2.33)(0.838) = 21.95 \text{ yards}$$

Because our sample mean is 22.5 yards, this falls within the "Reject H_0" region as shown in Figure 15.7. Our conclusion is that we have enough evidence to support the hypothesis that the mean increase in distance off the tee with my new balls exceeds 20 yards. I'm in business!

Figure 15.7

Hypothesis test for the golf ball example (original variable scale).

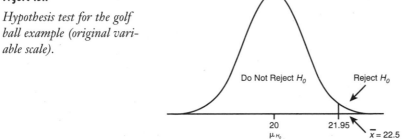

Mean Increase in Distance off the Tee in Yards

Random Thoughts

You might be asking yourself, "If the sample mean was 21.0 yards, shouldn't that provide conclusive evidence that my new ball increases distance by more than 20 yards?" According to the previous figure, the answer is no. Because we are basing our decision on a sample, an average of 21 is just too close to 20 to satisfy my claim. The sample average would have to be 21.95 yards or more in order to reject the null hypothesis.

We can also investigate this hypothesis test using the standardized scale. First we calculate the z-score that corresponds to our sample mean of 22.5 yards as follows:

$$z = \frac{\bar{x} - \mu_{H_0}}{\sigma_{\bar{x}}} = \frac{22.5 - 20}{0.838} = +2.98$$

As you can see in Figure 15.8, the calculated z-score is within the "Reject H_0" region and is consistent with our previous findings.

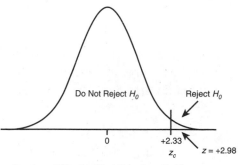

Do Not Reject H_0 Reject H_0

0   +2.33 $z = +2.98$
z_c

Number of Standard Deviations from the Mean

Figure 15.8

Hypothesis test for the golf ball example (standardized scale).

We will revisit this example in Chapter 17 and demonstrate how to perform a hypothesis test when comparing the mean of two populations with a dependent sample. Stay tuned!

The following table summarizes the critical z-scores for various levels of α.

Bob's Basics _____

It is helpful advice to draw the sampling distribution along with the rejection regions when performing hypothesis tests. A picture is worth a thousand words even in statistics!

Alpha	Tail	Critical z-Score
0.01	One	±2.33
0.01	Two	±2.57
0.02	One	±2.05
0.02	Two	±2.33
0.05	One	±1.64
0.05	Two	±1.96
0.10	One	±1.28
0.10	Two	±1.64

Also note that the critical z-scores for the right tail of the sampling distribution are always a positive value and the critical z-scores for the left tail of the sampling distribution are always a negative value.

As I mentioned earlier, the purpose of this chapter was to introduce the basic concepts of hypothesis testing. The following two chapters will go on to explore hypothesis testing in even more loving detail. So hang in there—we're just getting warmed up!

Your Turn

1. Formulate a hypothesis statement for the following claim: "The average adult drinks 1.7 cups of coffee per day." A sample of 35 adults drank an average of 1.95 cups per day. Assume the population standard deviation is 0.5 cups. Using $\alpha = 0.10$, test your hypothesis. What is your conclusion?

2. Formulate a hypothesis statement for the following claim: "The average age of our customers is less than 40 years old." A sample of 50 customers had a average age of 38.7 years. Assume the population standard deviation is 12.5 years. Using $\alpha = 0.05$, test your hypothesis. What is your conclusion?

3. Formulate a hypothesis statement for the following claim: "The average life of our light bulbs is more than 1,000 hours." A sample of 32 light bulbs had an average life of 1,190 hours. Assume the population standard deviation is 325 hours. Using $\alpha = 0.02$, test your hypothesis. What is your conclusion?

4. Formulate a hypothesis statement for the following claim: "The average delivery time is less than 30 minutes." A sample of 42 deliveries had an average time of 26.9 minutes. Assume the population standard deviation is 8 minutes. Using $\alpha = 0.01$, test your hypothesis. What is your conclusion?

The Least You Need to Know

♦ The null hypothesis, denoted by H_0, represents the status quo and involves stating the belief that the mean of the population is $\leq$, $=$, or $\geq$ a specific value.

♦ The alternative hypothesis, denoted by H_1, represents the opposite of the null hypothesis and holds true if the null hypothesis is found to be false.

♦ A two-tail hypothesis test is used whenever the alternative hypothesis is expressed as $\neq$ whereas a one-tail hypothesis test involves the alternative hypothesis being stated as $<$ or $>$.

♦ A Type I error occurs when the null hypothesis is rejected when, in reality, it is true. The probability of this error occurring is known as α, the level of significance.

♦ A Type II error occurs when the null hypothesis is accepted when, in reality, it is not true. The probability of this error occurring is known as β, the power of the hypothesis test.

Hypothesis Testing with One Sample

In This Chapter

◆ Testing the mean of a population using a large and small sample

◆ Examining the role of alpha (α) in hypothesis testing

◆ Using the *p*-value to test a hypothesis

◆ Testing the proportion of a population using a large sample

In Chapter 15, I introduced the concept of hypothesis testing to whet your appetite. This chapter is devoted to hypothesis testing that involves only one population, whereas Chapter 17 will discuss testing that compares two different populations to each other.

Hypothesis testing involving one population focuses on confirming claims such as the population average is equal to a specific value. There are many different cases to be considered with this type of hypothesis testing that will be examined in the following sections. This chapter relies on many of the concepts that were explored in Chapters 14 and 15, so you might want to be sure you are comfortable with that material before diving into this chapter.

Hypothesis Testing for the Mean with Large Samples

When the sample size that we use to test our hypothesis is large ($n \geq 30$), we can rely on our old friend the central limit theorem that we met in Chapter 13. However, we still have 2 cases to consider; whether σ, the population standard deviation, is known or unknown.

When Sigma Is Known

To demonstrate this type of hypothesis test, I'll use the following story.

One of the most feared phrases a husband can hear from his wife is, "Honey, let's go on a diet together." I should have been suspicious of Deb's motives when she suggested we go on the low-carbohydrate diet, especially because she wears size 2 pants. But I guess I could stand to lose a few pounds, so in a weak moment, I agreed. After all, I figured we could turn this into a competition to make things more interesting.

After a few harrowing days without my beloved carbohydrates (who would have guessed a grown man could dream about Cheez-its night after night), I began to wonder how Deb was doing so well with the diet. I found the answer to this mystery hidden deep in the trunk of her car—a half-eaten box of *cinnamon rolls*. I guess that makes me the winner. The thrill of victory!

Anyway, let's say that this particular diet claims that the average age of the person who participates in this self-inflicted torture is less than 40 years old. We set up our hypothesis as follows:

$H_0 : \mu \geq 40$ years old

$H_1 : \mu < 40$ years old

We sample 60 people on the diet and find that their average age is 35.7 years. Given that σ, the population standard deviation, is 16 years, we'll test the hypothesis at $\alpha = 0.05$.

Bob's Basics

Remember from Chapter 15 that α, the level of significance, represents the probability of making a Type I error. A Type I error occurs when we reject H_0, when H_0 is actually true. In this case, a Type I error would mean that we believe the claim that the average person on the diet is less than 40 years old when, in reality, the claim is not true. For this example there's a 5 percent chance of this error happening.

Because the sample size is greater than 30 and we know the value of σ, we calculate the z-score from the standardized normal distribution as we did in Chapter 15.

$$z = \frac{\bar{x} - \mu_{H_0}}{\sigma_{\bar{x}}}$$

For our example, the standard error of the mean, $\sigma_{\bar{x}}$, would be:

$$\sigma_{\bar{x}} = \frac{\sigma}{\sqrt{n}} = \frac{16}{\sqrt{60}} = 2.07 \text{ years}$$

This results in a calculated z-score of:

$$z = \frac{\bar{x} - \mu_{H_0}}{\sigma_{\bar{x}}} = \frac{35.7 - 40}{2.07} = -2.08$$

Also recall from Chapter 15, the critical z-score, which defines the boundary for the rejection region, is –1.64 for a one-tail (left side) test with $\alpha = 0.05$. Figure 16.1 shows this test graphically.

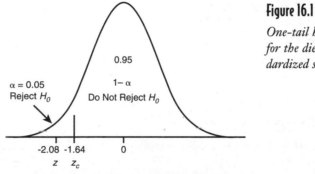

Figure 16.1

One-tail hypothesis test for the diet example (standardized scale).

0.95

$\alpha = 0.05$
Reject H_0

$1 - \alpha$
Do Not Reject H_0

-2.08 -1.64   0
 z z_c

Number of Standard Deviations from the Mean

As you can see in the figure, the calculated z-score of –2.08 falls within the "Reject H_0" region, which allows us to conclude that the claim that the average age of those on this diet is less than 40 years old. I knew I was too old for this diet! In general, we reject H_0 if $|z| > |z_c|$, where $|z|$ means the "absolute value of z." For instance, $|-2.08| = 2.08$.

When Sigma Is Unknown

Many times, we just don't have enough information to know the value of σ, the population standard deviation. However, as long as our sample size is 30 or more, we can substitute s, the sample standard deviation for σ. To illustrate this technique, we'll use the following example.

I don't know about you, but it seems I spend too much time on the phone waiting on hold for a live customer service representative. Let's say a particular company has claimed that the average time a customer waits on hold is less than 5 minutes. We'll assume we do not know the value of σ. The following table represents the wait time in minutes for a random sample of 30 customers.

Wait Time in Minutes

6.2	3.8	1.3	5.4	4.7	4.4	4.6	5.0	6.6	8.3
3.2	2.7	4.0	7.3	3.6	4.9	0.5	2.9	2.5	5.6
5.5	4.7	6.5	7.1	4.4	5.2	6.1	7.4	4.8	2.9

Using Excel, we can determine that $\bar{x} = 4.74$ minutes and $s = 1.82$ minutes. At first glance, it appears the company's claim is valid. But let's put it through a hypothesis test with $\alpha = 0.02$ to be sure.

The hypothesis can be stated as:

$H_0: \mu \geq 5.0$ minutes

$H_1: \mu < 5.0$ minutes

From Chapter 15, we know that the critical z-score for a one-tail (left side) hypothesis test with $\alpha = 0.02$ is -2.05.

As we did earlier in Chapter 14, we can approximate the standard error of the mean by:

$$\sigma^{\mu}_{x} = \frac{s}{\sqrt{n}} = \frac{1.82}{\sqrt{30}} = 0.332 \text{ minutes}$$

Our calculated z-score using this particular sample would be:

$$z = \frac{\bar{x} - \mu_{H_0}}{\sigma^{\mu}_{x}} = \frac{4.74 - 5.0}{0.332} = -0.78$$

Figure 16.2 shows this test graphically.

According to our figure, we do not reject the null hypothesis. In other words, we do not have enough evidence from this sample to support the company's claim that the average wait on hold is less than 5 minutes. Even though the sample average is actually less than 5 minutes (4.74), it is too close to 5 minutes to say there is a difference between the 2 values. Another way to state this is to say: "The difference between 4.74 and 5.0 is not statistically significant in this case."

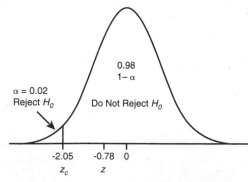

The Role of Alpha in Hypothesis Testing

For all of the examples in these last two chapters, I have just stated a value for α, the level of significance. I'm sure you're wondering to yourself what impact will changing the value of α have on the hypothesis test. Great question!

Suppose that I am making a claim that the average grade for a person using this book will be more than an 87. (I'm not *really* making this claim, so don't get too excited!) The hypothesis test would be stated as follows:

$$H_0 : \mu \le 87$$

$$H_1 : \mu > 87$$

Now, it would be in my best interest if I could reject H_0, which would validate my claim. I can do so by choosing a fairly high value for α, say 0.10. This corresponds to a critical z-score of +1.28, because we are using the right tail of a one-tail hypothesis test.

Let's say that σ, the population standard deviation, is 12 and my sample mean is 90.6, which was taken from a sample size of 32 students. For this example, the standard error of the mean, $\sigma_{\bar{x}}$, would be:

$$\sigma_{\bar{x}} = \frac{\sigma}{\sqrt{n}} = \frac{12}{\sqrt{32}} = 2.12$$

This results in a calculated z-score of:

$$z = \frac{\bar{x} - \mu_{H_0}}{\sigma_{\bar{x}}} = \frac{90.6 - 87}{2.12} = +1.70$$

According to Figure 16.3, I have achieved my goal of rejecting H_0, because the calculated z-score is within the shaded region. My book appears to have done the trick!

Figure 16.3

Hypothesis test for grade example, α = 0.10.

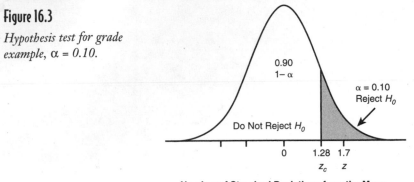

However, I must admit, I chose a pretty "wimpy" value of α = 0.10 in an effort to help prove my claim. In this case, I am willing to accept a 10 percent chance of a Type I error. A more impressive test would be to set alpha lower, say α = 0.01. Now that's a "real man's alpha." The level of significance corresponds to a critical z-score of +2.33. Figure 16.4 shows the impact of this change.

Figure 16.4

Hypothesis test for grade example, α = 0.01.

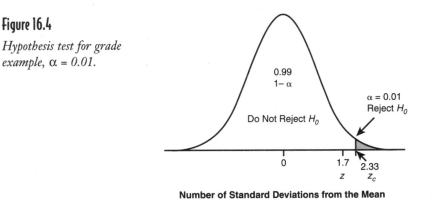

As you can see, to my horror, the shaded region no longer includes my calculated z-score of +1.7. Therefore, I do not reject H_0 and cannot claim the average grade of those using my book exceeds an 87. In general, a hypothesis test that rejects H_0 is most impressive with a low value of α.

Introducing the *p*-Value

Just when you thought it was safe to get back in the water, along comes another shark. This is the perfect opportunity to throw another concept at you. You might feel like grumbling a little right now, but in the end you'll be thanking me.

The *p-value* is the smallest level of significance at which the null hypothesis will be rejected, assuming the null hypothesis is true. The *p*-value is sometimes referred to as the *observed level of significance*. I know this may sound like a lot of mumbo-jumbo right now, but an illustration will help make this clear.

Stat Facts

The **observed level of significance** is the smallest level of significance at which the null hypothesis will be rejected, assuming the null hypothesis is true. It is also known as the *p*-value.

The *p*-Value for a One-Tail Test

Using the previous grade example (over 87 if using this book), the *p*-value is represented by the shaded area to the right of the calculated z-score of +1.7. This is shown in Figure 16.5.

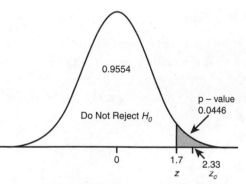

Figure 16.5

p-*value for the grade example.*

Using our standardized normal z table (Table 3 in Appendix B), we can confirm that the shaded area in the right tail is equal to $P[z > +1.7] = 0.0446$.

Bob's Basics

Recall that $P[z > +1.7] = 1 - P[z \le +1.7] = 1 - 0.9554 = 0.0446$. See Chapter 11 if you need a refresher on using the standardized normal z table.

Because our *p*-value of 0.0446 is more than the value of α (set at 0.01), we do not reject H_0. Most statistical software packages (including Excel) provide *p*-values with the analysis.

Another way to describe this *p*-value is to say, in a very scholarly voice, "our results are significant at the 0.0446 level." This means that as long as the value of α is 0.0446 or larger, we will reject H_0, which is normally good news for researchers trying to validate their findings.

Stat Facts

We can use the **p-value** to determine whether or not to reject the null hypothesis. In general …

◆ If *p*-value ≤ α, we reject the null hypothesis.

◆ If *p*-value > α, we do not reject the null hypothesis.

Calculating the *p*-value for a two-tail hypothesis test is slightly different and is shown in the next section.

The *p*-Value for a Two-Tail Test

Recall that a two-tail hypothesis test is used when the null hypothesis is stated as an equality. For example, let's test a claim that states the average number of miles driven by a passenger vehicle in a year equals 11,500 miles. I have serious reservations about this claim after spending half the day being a taxi driver to the kids. The hypotheses would be stated as follows:

$H_0 : \mu = 11,500$ miles

$H_1 : \mu \neq 11,500$ miles

Let's assume σ = 3,000 miles and we want to set α = 0.05. We sample 80 drivers and determine the average number of miles driven is 11,900. What is our *p*-value and what do we conclude about the hypothesis?

For this example, the standard error of the mean, $\sigma_{\bar{x}}$, would be:

$$\sigma_{\bar{x}} = \frac{\sigma}{\sqrt{n}} = \frac{3000}{\sqrt{80}} = 335.41 \text{ miles}$$

This results in a calculated z-score of:

$$z = \frac{\bar{x} - \mu_{H_0}}{\sigma_{\bar{x}}} = \frac{11,900 - 11,500}{335.41} = +1.19$$

The critical z-score for a two-tail test with $\alpha = 0.05$ is ± 1.96. The shaded area in Figure 16.6 shows the p-value for this test.

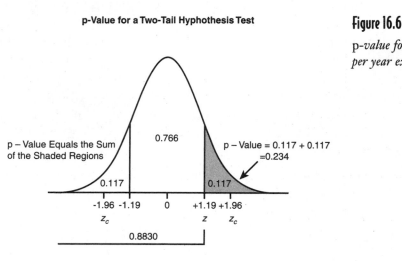

p-Value for a Two-Tail Hyphothesis Test

Figure 16.6

p-value for the miles driven per year example.

According to Table 3 in Appendix B, the $P[z \leq +1.19] = 0.8830$. This means the shaded region in the right tail of Figure 16.6 is $P[z > +1.19] = 1 - 0.8830 = 0.117$. Because this is a two-tail test, we need to double this area to arrive at our p-value. According to our figure, the p-value is the total area of both shaded regions, which is $2 \times 0.117 = 0.234$. Because $p > a$, we do not reject the null hypothesis. Our data supports the claim that the average number of miles driven per year by a passenger vehicle is 11,500.

In general, the smaller the p-value, the more confident we are about rejecting the null hypothesis. In most cases a researcher is attempting to find support for the alternative hypothesis. A low p-value provides support that brings joy to his or her heart.

Hypothesis Testing for the Mean with Small Samples

Recall from Chapter 14, with a small sample size, we lose the use of the central limit theorem and, therefore, need to assume that the population is normally distributed for all cases in this section. The first case that we'll examine is when we know σ, the population standard deviation.

When Sigma Is Known

When σ is known, the hypothesis test reverts back to the large sample size case. We can do this because we are now assuming the population is normally distributed. We can demonstrate this method with the following example.

Opening up my monthly cell phone bill lately has become a nerve-wracking experience. As I warily open the envelope, I wonder what surprises await me. With several users on our family "share plan," I can often count on somebody discovering a new feature that has nothing to do with talking to another person on the phone and using this new-found discovery over and over and over again. Occasionally, after digging through countless pages full of numbers and codes, I breathe a sigh of relief and say a silent prayer of thanks. Most months, however, I end up clutching my chest and screaming "AIEEEEEEEE!" It's like playing a subtle form of Russian roulette with the phone company.

Anyway, let's say the phone company claims that the average monthly cell phone bill for their customers is $92 (I wish). We can test this claim by stating our hypothesis as:

$H_0: \mu = \$92$

$H_1: \mu \neq \$92$

We'll assume that σ = $22.50 and that the population is normally distributed. We select 18 phone bills randomly and determine the sample average equals $107. Using σ = 0.02, what do we conclude?

Bob's Basics

Recall from Chapter 14 that because we know σ and we assumed the population is normally distributed, we can use the z-scores from the normal probability distribution to test this hypothesis.

For this example, the standard error of the mean, $\sigma_{\bar{x}}$, would be:

$$\sigma_{\bar{x}} = \frac{\sigma}{\sqrt{n}} = \frac{\$22.50}{\sqrt{18}} = \$5.30$$

This results in a calculated z-score of:

$$z = \frac{\bar{x} - \mu_{H_0}}{\sigma_{\bar{x}}} = \frac{\$107 - \$92}{\$5.30} = +2.83$$

The critical z-score for a two-tail test with α = 0.02 is ±2.33. Figure 16.7 shows this test graphically.

As you can see in Figure 16.7, the calculated z-score of +2.83 is with the "Reject H_0" region. We therefore conclude that the average cell phone bill is not equal to $92. I didn't think so!

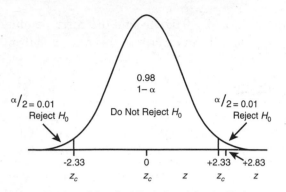

Figure 16.7

Hypothesis test for cell phone bills.

When Sigma Is Unknown

As we did in Chapter 14, when σ is unknown for a small sample size taken from a normally distributed population, we use the Student's t-distribution. This particular distribution allows us to substitute *s*, the sample standard deviation for σ.

As an example, suppose my son John claims his average golf score is less than 88. Not to be one to doubt him, I can test this claim with the following hypothesis:

$$H_0 : \mu \geq 88$$

$$H_1 : \mu < 88$$

We will assume that we do not know σ and that John's scores follow a normal distribution. The following represents a random sample of 10 golf scores from John.

John's Golf Scores

86   87   85   90   86   84   84   91   87   83

Using Excel, we can determine that $\bar{x}$ = 86.3 and s = 2.58 for this sample. Recall from Chapter 14, we can approximate the standard error of the mean using the following equation:

$$\sigma^{\mu}_{\bar{x}} = \frac{s}{\sqrt{n}} = \frac{2.58}{\sqrt{10}} = 0.816$$

We can then determine the calculated t-score using the following equation:

$$t = \frac{\bar{x} - \mu_{H_0}}{\hat{\sigma}_{\bar{x}}} = \frac{86.3 - 88}{0.816} = -2.08$$

We'll test this hypothesis using α = 0.05. To find the corresponding critical t-score, we use Table 4 from Appendix B. An excerpt of this table is shown here.

Student's t-Distribution Table

Selected right tail areas with confidence levels underneath

conf lev	0.2000	0.1500	0.1000	0.0500	0.0250	0.0100	0.0050	0.0010	0.0005
df	0.6000	0.7000	0.8000	0.9000	0.9500	0.9800	0.9900	0.9980	0.9990
1	1.376	1.963	3.078	6.314	12.706	31.821	63.657	318.31	636.62
2	1.061	1.386	1.886	2.920	4.303	6.965	9.925	22.327	31.599
3	0.978	1.250	1.638	2.353	3.182	4.541	5.841	10.215	12.924
4	0.941	1.190	1.533	2.132	2.776	3.747	4.604	7.173	8.610
5	0.920	1.156	1.476	2.015	2.571	3.365	4.032	5.893	6.869
6	0.906	1.134	1.440	1.943	2.447	3.143	3.707	5.208	5.959
7	0.896	1.119	1.415	1.895	2.365	2.998	3.499	4.785	5.408
8	0.889	1.108	1.397	1.860	2.306	2.896	3.355	4.501	5.041
9	0.883	1.100	1.383	1.833	2.262	2.821	3.250	4.297	4.781
10	0.879	1.093	1.372	1.812	2.228	2.764	3.169	4.144	4.587

Recall from Chapter 14, we need to determine the number of degrees of freedom, which is equal to $n - 1 = 10 - 1 = 9$ for this example. Because this is a one-tail (left side) test, we look under the $\alpha = 0.05$ column resulting in a critical t-score, t_c, equal to -1.833, which is underlined. This test is shown graphically in Figure 16.8.

Figure 16.8

Hypothesis test for John's golf scores.

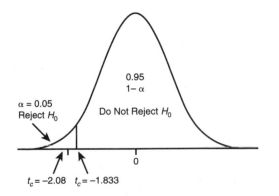

As we can see in the figure, the calculated t-score of –2.08 falls within the shaded "Reject H_0" region. Therefore, we can conclude that John's average golf score is indeed lower than 88. So that explains why he usually beats me! In general, we reject H_0 if $|t| > |t_c|$.

Let's take another example to demonstrate a two-tail hypothesis test using the t-distribution. I would like to test a claim that the average speed of cars passing a specific spot on the interstate is 65 miles per hour. We can express the hypothesis test as follows:

$H_0: \mu = 65$ miles per hour

$H_1: \mu \neq 65$ miles per hour

We will assume that we do not know σ and that speeds follow a normal distribution. The following represents a random sample of the speed of 7 cars.

Car Speeds

62   74   65   68   71   64   68

Stat Facts

Because John's golf score example is a one-tail test on the left side of the distribution, we use a negative critical t-score. Had this been a one-tail test on the right side, we would use a positive critical t-score.

Bob's Basics

It is not possible to determine the *p*-value for a hypothesis test when using the Student's t-distribution table in Appendix B. However, most statistical software will provide the *p*-value as part of the standard analysis. We'll see this in later chapters as we use Excel.

Using Excel, we can determine that $\bar{x} = 66.9$ mph and $s = 4.16$ mph for this sample. We can approximate the standard error of the mean:

$$\sigma^{\mu}_{\bar{x}} = \frac{s}{\sqrt{n}} = \frac{4.16}{\sqrt{7}} = 1.57 \text{ mph}$$

We can then determine the calculated t-score:

$$t = \frac{\bar{x} - \mu_{H_0}}{\sigma^{\mu}_{\bar{x}}} = \frac{66.9 - 65}{1.57} = +1.21$$

We'll test this hypothesis using $\alpha = 0.05$. To find the corresponding critical t-score, we use Table 4 from Appendix B. An excerpt of this table is shown here.

Student's t-Distribution Table

Selected right tail areas with confidence levels underneath

conf lev df	0.2000 0.6000	0.1500 0.7000	0.1000 0.8000	0.0500 0.9000	0.0250 0.9500	0.0100 0.9800	0.0050 0.9900	0.0010 0.9980	0.0005 0.9990
1	1.376	1.963	3.078	6.314	12.706	31.821	63.657	318.31	636.62
2	1.061	1.386	1.886	2.920	4.303	6.965	9.925	22.327	31.599
3	0.978	1.250	1.638	2.353	3.182	4.541	5.841	10.215	12.924
4	0.941	1.190	1.533	2.132	2.776	3.747	4.604	7.173	8.610
5	0.920	1.156	1.476	2.015	2.571	3.365	4.032	5.893	6.869
6	0.906	1.134	1.440	1.943	2.447	3.143	3.707	5.208	5.959
7	0.896	1.119	1.415	1.895	2.365	2.998	3.499	4.785	5.408

The number of degrees of freedom for this example equals $n - 1 = 7 - 1 = 6$. Because this is a two-tail test, we need to divide $\alpha = 0.05$ into 2 equal portions, 1 on the right side of the distribution, the other on the left. We then look under the $\alpha/2 = 0.025$ column resulting in a critical t-score, t_c, equal to ±2.447, which is underlined. This test is shown graphically in Figure 16.9.

Figure 16.9

Hypothesis test for car speed.

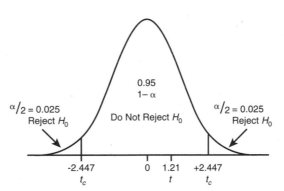

Hypothesis Test for Car Speeds
(Two Tail t-Distribution)

0.95
$1 - \alpha$

$\alpha/2 = 0.025$
Reject H_0

Do Not Reject H_0

$\alpha/2 = 0.025$
Reject H_0

-2.447   0   1.21   +2.447
t_c t t_c

As we can see in the figure, the calculated t-score of +1.21 falls within the "Do Not Reject H_0" region. Therefore, we can conclude that the average speed past this spot of the interstate averages 65 miles per hour.

Using Excel's TINV Function

We can generate critical t-scores using Excel's TINV function, which has the following characteristics:

TINV(probability, deg_freedom)

where:

probability = the level of significance, α, for a two-tail test

deg_freedom = the number of degrees of freedom

For instance, Figure 16.10 shows the TINV function being used to determine the critical t-score for $\alpha = 0.05$ and *d.f.* = 6 from our previous example, which was a two-tail test.

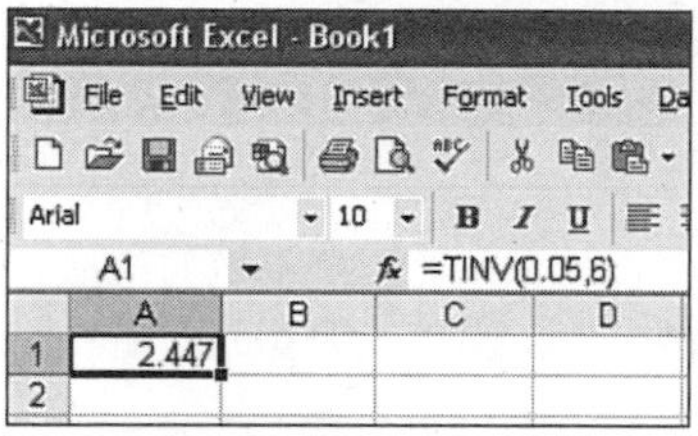

Figure 16.10

Excel's TINV function for a two-tail test.

Cell A1 contains the Excel formula =TINV(0.05, 6) with the result being 2.447. This probability is underlined in the previous table.

A one-tail test requires a slight modification. We need to multiply the probability in the TINV function by two because this parameter is based on a two-tail test. Figure 16.11 shows the TINV function being used to determine the critical t-score for $\alpha = 0.05$ and *d.f.* = 9 from our earlier one-tail test example with John's golf scores.

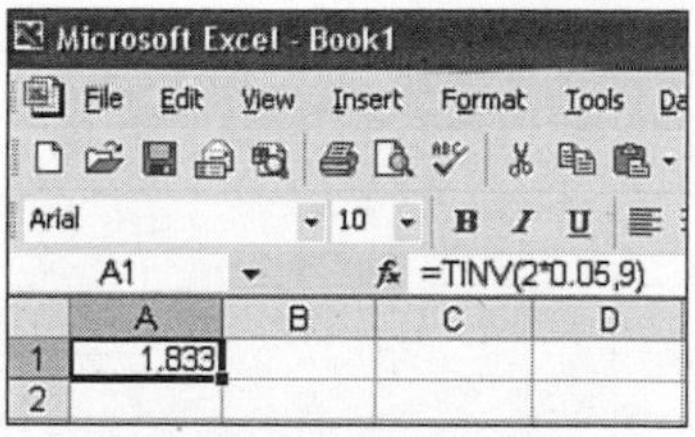

Figure 16.11

Excel's TINV function for a one-tail test.

Cell A1 contains the Excel formula =TINV(2*0.05, 9) with the result being 1.833. This is consistent with the result from our previous example.

Hypothesis Testing for the Proportion with Large Samples

Hypothesis testing can be performed for the proportion of a population as long as the sample size is large enough. Recall from Chapter 13, proportion data follows the binomial distribution, which can be approximated by the normal distribution under the following conditions:

$np \geq 5$ and $nq \geq 5$

where:

p = the probability of a success in the population

q = the probability of a failure in the population ($q = 1 - p$)

We will examine both one-tail and two-tail hypothesis testing for the proportion in the following sections.

One-Tail Hypothesis Test for the Proportion

Let's say we would like to test the hypothesis that more than 30 percent of U.S. households have Internet access. The hypothesis would be stated as:

$H_0: p \leq 0.30$

$H_1: p > 0.30$

where p = the proportion of U.S. households with Internet access.

Wrong Number

Be careful not to confuse this definition of p with the p-value that we talked about earlier.

We collect a sample of 150 households and find that 38 percent of these have Internet access. What can we conclude at the $\alpha = 0.05$ level?

Our first step is to calculate σ_p, the standard error of the proportion, which was described in Chapter 13 using the following equation:

$$\sigma_p = \sqrt{\frac{p_{H_0}\left(1 - p_{H_0}\right)}{n}}$$

where p_{H_0} = the proportion assumed by the null hypothesis. For our example:

$$\sigma_p = \sqrt{\frac{(0.30)(1 - 0.30)}{150}} = 0.037$$

Next, we can determine the calculated z-score using:

$$z = \frac{\bar{p} - p_{H_0}}{\sigma_p}$$

where $\bar{p}$ = the sample proportion. For our example:

$$z = \frac{\bar{p} - p_{H_0}}{\sigma_p} = \frac{0.38 - 0.30}{0.037} = +2.16$$

The critical z-score for a one-tail test with $\alpha = 0.05$ is +1.64. This hypothesis test is shown graphically in Figure 16.12.

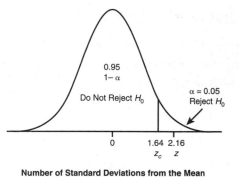

Figure 16.12

Hypothesis test for the Internet access example.

As you can see in Figure 16.12, the calculated z-score of +2.16 is within the "Reject H_0" region. Therefore, we conclude that the proportion of U.S. households with Internet access exceeds 30 percent.

The *p*-value for this test can be shown graphically in Figure 16.13.

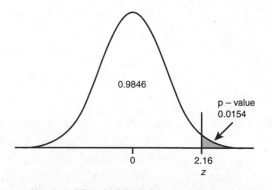

Figure 16.13

p-*value for the Internet access example.*

Using our standardized normal z table (Table 3 in Appendix B), we can confirm that the shaded area in the right tail is equal to:

$$P[z > +2.16] = 1 - P[z \leq +2.16]$$

$$P[z > +2.16] = 1 - 0.9846 = 0.0154$$

Therefore, our results are significant at the 0.0154 level. As long as $\alpha \geq 0.0154$, we will be able to reject H_0.

Two-Tail Hypothesis Test for the Proportion

We'll wrap this chapter up with one final two-tail example. Here, we want to test a hypothesis for a company that claims 50 percent of their customers are of the male gender. We state our hypothesis as:

$$H_0 : p = 0.50$$

$$H_1 : p \neq 0.50$$

We randomly select 200 customers and find that 47 percent are male. What can we conclude at the $\alpha = 0.05$ level?

We need to determine σ_p, the standard error of the proportion:

$$\sigma_p = \sqrt{\frac{p_{H_0}(1 - p_{H_0})}{n}} = \sqrt{\frac{(0.50)(1 - 0.50)}{200}} = 0.035$$

Next, we can determine the calculated z-score:

$$z = \frac{\overline{p} - p_{H_0}}{\sigma_p} = \frac{0.47 - 0.50}{0.035} = -0.86$$

Bob's Basics

In general, we reject H_0 if $|z| > |z_c|$ or $|t| > |t_c|$. Also, we do not reject H_0 if $|z| \leq |z_c|$ or $|t| \leq |t_c|$.

The critical z-score for a two-tail test with $\alpha = 0.05$ is ±1.96. This hypothesis test is shown graphically in Figure 16.14.

As you can see in Figure 16.14, the calculated z-score of -0.86 is within the "Do Not Reject H_0" region. There, we conclude that the proportion of male customers is equal to 50 percent for this company.

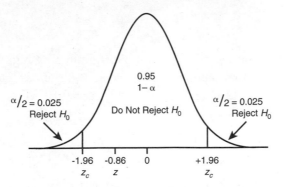

Figure 16.14

Hypothesis test for the percentage of males example.

The *p*-value for this test can be shown graphically in Figure 16.15.

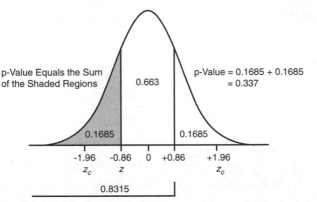

Figure 16.15

p-*value for the percentage of males example.*

Using our standardized normal z table (Table 3 in Appendix B), we can confirm that the shaded area in the left tail is equal to:

$$P[z \leq -0.86] = 1 - P[z \leq +0.86]$$

$$P[z \leq -0.86] = 1 - 0.8315 = 0.1685$$

Because this is a two-tail test, the *p*-value would be $2 \times 0.1685 = 0.337$, which represents the total area in both shaded regions.

Your Turn

1. Test the claim that the average SAT score for graduating high school students is equal to 1100. A random sample of 70 students was selected, and the average SAT score was 1,035. Assume $\sigma = 310$ and use $\alpha = 0.10$. What is the p-value for this sample?

2. A small business college claims that the average class size is equal to 35 students. Test this claim at $\alpha = 0.02$ using the following sample of class size:

   42   28   36   47   35   41   33   30   39   48

   Assume the population is normally distributed and that σ is unknown.

3. Test the claim that the average gasoline consumption per car in the United States is more than 7 liters per day. (We're going metric here!) Use the random sample here, which represents daily gasoline usage for one car:

   9   6   4   12   4   3   18   10   4   5

   3   8   4   11   3   5   8   4   12   10

   9   5   15   17   6   13   7   8   14   9

   Assume the population is normally distributed, and that σ is unknown. Use $\alpha = 0.05$ and determine the p-value for this sample.

4. Test the claim that the proportion of Republican voters in a particular city is less than 40 percent. A random sample of 175 voters was selected and found to consist of 30 percent Republicans. Use $\alpha = 0.01$ and determine the p-value for this sample.

The Least You Need to Know

   ◆ The smaller the value of α, the level of significance, the more difficult it is to reject the null hypothesis.

   ◆ We reject H_0 if $|z| > |z_c|$ or $|t| > |t_c|$.

   ◆ The p-value is the smallest level of significance at which the null hypothesis will be rejected, assuming the null hypothesis is true.

   ◆ If the p-value $\leq \alpha$, we reject the null hypothesis. If p-value $> \alpha$, we do not reject the null hypothesis.

   ◆ The Student's t-distribution is used for the hypothesis test when $n < 30$, σ is unknown, and the population is normally distributed.

Chapter 17

Hypothesis Testing with Two Samples

In This Chapter

◆ Developing the sampling distribution for the difference in means

◆ Testing the difference in means between populations using a large and small sample

◆ Distinguishing between independent and dependent samples

◆ Using Excel to perform a hypothesis test

◆ Testing the difference in proportions between populations

Now we're really cooking. Because you have done so well with one sample hypothesis testing, you are ready to graduate to the next level—two-sample testing. Here we are often testing to see whether there is a difference between two separate populations. For instance, I could test to see whether there was a difference between Brian and John's average golf score. But being an "experienced" parent, I know better than to go near that one.

Because there are many similarities between the concepts of this chapter and Chapter 16, it would be a good idea to have a firm handle on the previous chapter before diving into this one.

The Concept of Testing Two Populations

Many statistical studies involve comparing the some parameter, such as a mean, between two different populations. For example:

◆ Is there a difference in average SAT scores between males and females?

◆ Do "long-life" light bulbs really outlast standard light bulbs?

Stat Facts

The **sampling distribution for the difference in means** describes the probability of observing various intervals for the difference between 2 sample means.

◆ Does the average selling price of a house in Newark differ from the average selling price for a house in Wilmington?

To answer such questions, we need to explore a new sampling distribution. (I promise this will be the last.) This one has the fanciest name of them all—the *sampling distribution for the difference in means*. (Dramatic background music brings us to the edge of our seats.)

Sampling Distribution for the Difference in Means

The sampling distribution for the difference in means can best be described in Figure 17.1.

Figure 17.1

The sampling distribution for the difference in means.

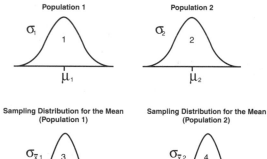

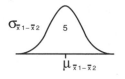

As an example, let's consider testing for a difference in SAT scores for male and female students. We'll assign female students as Population 1 and male students as Population 2. Graph 1 in Figure 17.1 represents the distribution of SAT scores for the female students with mean μ_1 and standard deviation σ_1. Graph 2 represents the same for the male population.

Graph 3 represents the sampling distribution for the mean for the female students. This graph is the result of taking samples of size n_1 and plotting the distribution of sample means. Recall that we discussed this distribution of sample means back in Chapter 13. The mean of this distribution would be:

$$\mu_{\bar{x}_1} = \mu_1$$

This is according to the central limit theorem from Chapter 13. The same logic holds true for Graph 4 for the male population.

Graph 5 in Figure 17.1 shows the distribution that represents the difference of sample means from the female and male populations. This is the sampling distribution for the difference in means, which has the following mean:

$$\mu_{\bar{x}_1 - \bar{x}_2} = \mu_{\bar{x}_1} - \mu_{\bar{x}_2}$$

In other words, the mean of this distribution, shown in Graph 5, is the difference between the means of Graphs 3 and 4.

The standard deviation for the Graph 5 is known as the *standard error of the difference between two means* and is calculated with:

$$\sigma_{\bar{x}_1 - \bar{x}_2} = \sqrt{\frac{\sigma_1^2}{n_1} + \frac{\sigma_2^2}{n_2}}$$

where:

σ_1^2, σ_2^2 = the variance for Populations 1 and 2

n_1, n_2 = the sample size from Populations 1 and 2

Now before you pull the rest of your hair out, let's put these guys to work in the following section.

Stat Facts

The **standard error of the difference between two means** describes the variation in the difference between 2 sample means and is calculated using

$$\sigma_{\bar{x}_1 - \bar{x}_2} = \sqrt{\frac{\sigma_1^2}{n_1} + \frac{\sigma_2^2}{n_2}}.$$

Testing for Differences Between Means with Large Sample Sizes

When the sample sizes from both populations of interest are greater than 30, the central limit theorem allows us to use the normal distribution to approximate the sampling distribution for the difference in means. We will demonstrate this technique with the following example.

Studies have been done to investigate the effects of stimulation on the brain development of rats. I guess the logic being what's good for rats can't be all that bad for us humans. Two samples were randomly selected from the same rat population.

The first sample, we'll call these the "lucky rats" (Population 1), were surrounded with every luxury a rat could imagine. I can envision a country club atmosphere, complete with a golf course (and tiny golf carts), tennis courts, and a five-star restaurant where our lucky rats could feast on imported cheese and French wine while they discuss the state of the rat economy.

The second sample, we'll call them the "less-fortunate rats" (Population 2), didn't have it quite so good. These guys were locked in a barren cage and were forced to eat Cheez Whiz from a can and watch reruns of reality TV shows. Animal rights activists protested against this experiment, claiming the involuntary use of Cheez Whiz was "inhumane."

After spending three months in each of these environments, the size of each rat brain was measured by weight for development. I'll spare you the details as to how this was done, but I will tell you that Harvey the Rat mysteriously failed to show for his 8 A.M. tee time. His group went off without him.

The following table summarizes these gruesome findings.

Summarized Data for Rat Experiment

Population	Average Brain Weight in Grams $\bar{x}$	Sample Standard Deviation s	Sample Size n
Lucky (1)	2.4	0.6	50
Less-Fortunate (2)	2.1	0.8	60

For this hypothesis test, we need to assume that the two samples are independent of each other. In other words, there is no relationship between the rats in the lucky sample and the rats in the less-fortunate sample. The hypothesis statement for this two-sample test would be as follows:

$$H_0: \mu_1 \le \mu_2$$

$$H_1: \mu_1 > \mu_2$$

where:

μ_1 = the mean brain weight of the lucky rat population

μ_2 = the mean brain weight of the less-fortunate rat population

The hypothesis can also be expressed as:

$$H_0: \mu_1 - \mu_2 \le 0$$

$$H_1: \mu_1 - \mu_2 > 0$$

The alternative hypothesis supports the claim that the lucky rats will have heavier brains. Seems to me this could lead to neck problems for these rats—but I'll leave that question for another study. We'll test this hypothesis at the $\alpha = 0.05$ level.

If σ_1 or σ_2 are not known, then we can use s_1 or s_2, the standard deviation from the samples of populations 1 and 2 as an approximation, as long as $n \ge 30$ for both populations, as shown here:

$$\sigma^{\mu} = s$$

With this assumption, we can approximate the standard error of the difference between 2 means using:

$$\sigma^{\mu}_{\bar{x}_1 - \bar{x}_2} = \sqrt{\frac{\sigma^{\mu 2}_1}{n_1} + \frac{\sigma^{\mu 2}_2}{n_2}}$$

Because we do not know σ_1 or σ_2 in our rat example, we set:

$$\sigma^{\mu}_1 = s_1 \text{ and } \sigma^{\mu}_2 = s_2$$

$$\sigma^{\mu}_{\bar{x}_1 - \bar{x}_2} = \sqrt{\frac{\sigma^{\mu 2}_1}{n_1} + \frac{\sigma^{\mu 2}_2}{n_2}} = \sqrt{\frac{(0.6)^2}{50} + \frac{(0.8)^2}{60}} = 0.134 \text{ grams}$$

We are now ready to determine the calculated z-score using the following equation:

$$z = \frac{(\bar{x}_1 - \bar{x}_2) - (\mu_1 - \mu_2)_{H_0}}{\sigma^{\mu}_{\bar{x}_1 - \bar{x}_2}}$$

Bob's Basics

The term $(\mu_1 - \mu_2)_{H_0}$ refers to the hypothesized difference between the two population means. When the null hypothesis is testing that there is no difference between population means, then the term $(\mu_1 - \mu_2)_{H_0}$ is set to 0.

For the rat example, our calculated z-score becomes:

$$z = \frac{(\bar{x}_1 - \bar{x}_2) - (\mu_1 - \mu_2)_{H_0}}{\sigma^{\mu}_{\bar{x}_1 - \bar{x}_2}} = \frac{(2.4 - 2.1) - 0}{0.134} = +2.24$$

The results of this hypothesis test are shown in Figure 17.2.

The critical z-score for a one-tail (right side) test with $\alpha = 0.05$ is +1.64. According to Figure 17.2, this places the calculated z-score of +2.24 in the "Reject H_0" region, which leads to our conclusion that the lucky rats have heavier brains than the less-fortunate rats.

Figure 17.2

Hypothesis test for rat example.

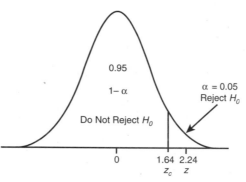

Number of Standard Deviations from the Mean

Stat Facts

The conditions that are necessary for the hypothesis test for differences between means with large sample sizes are as follows:

- The samples are independent of each other.
- The size of each sample must be greater than or equal to 30.
- If the population standard deviations are unknown, the sample standard deviations can be used to approximate them.

The *p*-value for this sample can be found using the normal z-score table found in Appendix B as follows:

$$P[z > +2.24] = 1 - P[z \leq +2.24]$$

$$P[z > +2.24] = 1 - 0.9875 = 0.0125$$

The results of our rat study can greatly improve the lives of many. When your spouse catches you sneaking off to the golf course on Saturday morning, you can tell him or her with a straight face that you are just trying to improve your mind. We now have the statistics to support you. But be warned, you might develop a sore neck with all that extra brain weight.

Testing a Difference Other Than Zero

In the previous example, we were just testing whether or not there was any difference between the two populations. We can also test whether the difference exceeds a certain value. As an example, suppose we want to test the hypothesis that the average salary of a mathematician in New Jersey exceeds the average salary in Virginia by more than $5,000. We would state the hypotheses as follows:

$$H_0 : \mu_1 - \mu_2 \leq 5,000$$

$$H_1 : \mu_1 - \mu_2 > 5,000$$

where:

μ_1 = the mean salary of a mathematician in New Jersey

μ_2 = the mean salary of a mathematician in Virginia

We'll assume that σ_1 = $8,100 and σ_2 = $7,600 and we'll test this hypothesis at the α = 0.10 level.

A sample of 42 mathematicians from New Jersey had a mean salary of $51,500, whereas a sample of 54 mathematicians from Virginia had a mean salary of $45,400.

The standard error of the difference between two means is:

$$\sigma_{\bar{x}_1 - \bar{x}_2} = \sqrt{\frac{\sigma_1^2}{n_1} + \frac{\sigma_2^2}{n_2}}$$

$$\sigma_{\bar{x}_1 - \bar{x}_2} = \sqrt{\frac{(8,100)^2}{42} + \frac{(7,600)^2}{54}} = \$1,622.3$$

Our calculated z-score becomes:

$$z = \frac{(\bar{x}_1 - \bar{x}_2) - (\mu_1 - \mu_2)}{\sigma_{\bar{x}_1 - \bar{x}_2}}$$

$$z = \frac{(\$51,500 - \$45,400) - (\$5,000)}{\$1,622.3} = +0.68$$

The results of this hypothesis test are shown in Figure 17.3.

Figure 17.3

Hypothesis test for the salary example.

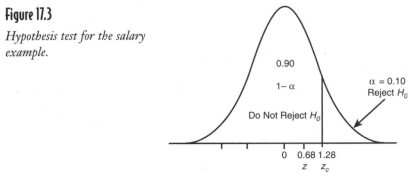

The critical z-score for a one-tail (right side) test with $\alpha = 0.10$ is $+1.28$. According to Figure 17.3, this places the calculated z-score of $+0.68$ in the "Do Not Reject H_0" region, which leads to our conclusion that the difference in salaries between the two states does not exceed $5,000.

Testing for Differences Between Means with Small Sample Sizes and Unknown Sigma

This section addresses the situation where the population standard deviation, σ, is not known and the sample sizes are small. If one or both of our sample sizes is less than 30, the population needs to be normally distributed to use any of the following techniques. We made the same assumption for small sample sizes back in Chapters 14 and 16.

The sampling distribution for the difference between sample means for this scenario follows the Student's t-distribution. Also for small sample sizes, the equation for the standard error of the difference between two means, $\sigma_{\bar{x}_1 - \bar{x}_2}$, depends on whether or not the standard deviations (or the variances) of the two populations are equal. The first example will deal with equal standard deviations.

Equal Population Standard Deviations

We have a very mysterious occurrence in our household—batteries seem to vanish into thin air. I started buying them in 24-packs at the warehouse store, naively thinking that "these will last a long time." Wrong again—the more I buy, the faster they disappear. Maybe it has something to do with certain teenagers listening to music on their portable CD players at a "brain-numbing" volume into the wee hours of the morning. Just a thought. So if I ever hear about a new "longer-lasting battery," I'm all over it. Let's say a company is promoting one of these batteries claiming that its life is significantly longer than regular batteries. The hypothesis statement would be:

$$H_0: \mu_1 \le \mu_2$$
$$H_1: \mu_1 > \mu_2$$

where:

μ_1 = the mean life of the long lasting batteries

μ_2 = the mean life of the regular batteries

We'll test this hypothesis at the $\alpha = 0.01$ level. The following data was collected measuring the battery life in hours for both types of batteries:

Raw Data for Battery Example

Long Lasting Battery (Population 1):
51   44   58   36   48   53   57   40   49   44   60   50

Regular Battery (Population 2):
42   29   51   38   39   44   35   40   48   45

Using Excel, we can summarize this data in the following table.

Summarized Battery Data

Population in Hours	Sample Mean $\bar{x}$	Sample Standard Deviation s	Sample Size n
Long-lasting (1)	49.2	6.40	12
Regular (2)	41.1	7.31	10

In this example, we are assuming that $\sigma_1 = \sigma_2$, but that the values of σ_1 and σ_2 are unknown. Under these conditions, we calculate a *pooled estimate of the standard deviation* using the following equation:

$$s_p = \sqrt{\frac{(n_1 - 1)s_1^2 + (n_2 - 1)s_2^2}{n_1 + n_2 - 2}}$$

Stat Facts

The **pooled estimate of the standard deviation** combines 2 sample variances into one variance and is calculated using $s_p = \sqrt{\dfrac{(n_1 - 1)s_1^2 + (n_2 - 1)s_2^2}{n_1 + n_2 - 2}}$.

Don't panic just yet. This equation looks a whole lot better with numbers plugged in.

$$s_p = \sqrt{\frac{(n_1 - 1)s_1^2 + (n_2 - 1)s_2^2}{n_1 + n_2 - 2}} = \sqrt{\frac{(12 - 1)(7.31)^2 + (10 - 1)(6.40)^2}{12 + 10 - 2}}$$

$$s_p = \sqrt{\frac{956.44}{20}} = 6.92$$

We can now approximate the standard error of the difference between two means using:

$$\sigma^{\mu}_{\bar{x}_1 - \bar{x}_2} = s_p \sqrt{\frac{1}{n_1} + \frac{1}{n_2}}$$

Let's apply our example to this fellow.

$$\sigma^{\mu}_{\bar{x}_1 - \bar{x}_2} = s_p \sqrt{\frac{1}{n_1} + \frac{1}{n_2}} = (6.92)\sqrt{\frac{1}{12} + \frac{1}{10}}$$

$$\sigma^{\mu}_{\bar{x}_1 - \bar{x}_2} = (6.92)\sqrt{0.1833} = 2.96 \ \text{hours}$$

We are now ready to determine our calculated t-score using the following equation:

$$t = \frac{(\bar{x}_1 - \bar{x}_2) - (\mu_1 - \mu_2)_{H_0}}{\sigma^{\mu}_{\bar{x}_1 - \bar{x}_2}} = \frac{(49.2 - 41.1) - 0}{2.96} = +2.73$$

The number of degrees of freedom for this test are:

$$d.f. = n_1 + n_2 - 2 = 12 + 10 - 2 = 20$$

The critical t-score, taken from Table 4 in Appendix B, for a one-tail (right) test using $\alpha = 0.10$ with *d.f.* = 20 is +2.528. This hypothesis test is shown graphically in Figure 17.4.

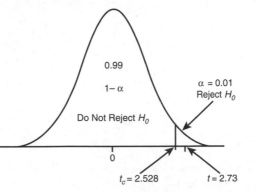

Figure 17.4

Hypothesis test for the battery example.

0.99

$1-\alpha$

Do Not Reject H_0

$\alpha = 0.01$
Reject H_0

0

$t_c = 2.528$

$t = 2.73$

According to Figure 17.4, our calculated t-score of +2.73 is found in the "Reject H_0" region, which leads to our conclusion that the long-lasting batteries do indeed have a longer life than the regular batteries. Now that has my attention.

Stat Facts

The conditions that are necessary for the hypothesis test for differences between means with small sample sizes are as follows:

♦ The samples are independent of each other.

♦ The population must be normally distributed.

♦ If σ_1 and σ_2 are known, use the normal distribution to determine the rejection region.

♦ If σ_1 and σ_2 are unknown, approximate them with s_1 and s_2 and use the Student's t-distribution to determine the rejection region.

This procedure was based on the assumption that the standard deviations of the populations were equal. What if this assumption is not true? I'm glad you asked!

Unequal Population Standard Deviations

We'll investigate this scenario using the same battery example, but now we will assume that $\sigma_1 \neq \sigma_2$. The procedure is identical to the previous method except for two changes.

The first difference involves the standard error of the difference between two means. The equation used for this scenario is as follows:

$$\sigma^{\mu}_{\bar{x}_1 - \bar{x}_2} = \sqrt{\frac{s_1^2}{n_1} + \frac{s_2^2}{n_2}}$$

For the battery example, our result is:

$$\sigma^{\mu}_{\bar{x}_1 - \bar{x}_2} = \sqrt{\frac{(7.31)^2}{12} + \frac{(6.40)^2}{10}} = \sqrt{(4.45) + (4.10)} = 2.92$$

We are now ready to determine our calculated t-score using the following equation:

$$t = \frac{(\bar{x}_1 - \bar{x}_2) - (\mu_1 - \mu_2)_{H_0}}{\sigma^{\mu}_{\bar{x}_1 - \bar{x}_2}} = \frac{(49.2 - 41.1) - 0}{2.92} = +2.77$$

The second difference (hold on to your hat) is the method for determining the number of degrees of freedom for the Student's t-distribution.

$$d.f. = \frac{\left(\frac{s_1^2}{n_1} + \frac{s_2^2}{n_2}\right)^2}{\frac{\left(\frac{s_1^2}{n_1}\right)^2}{n_1 - 1} + \frac{\left(\frac{s_2^2}{n_2}\right)^2}{n_2 - 1}}$$

Before you have a seizure, let me demonstrate that this animal's bark is worse than its bite. First recognize that for our battery example:

$$\frac{s_1^2}{n_1} = \frac{(7.31)^2}{12} = 4.45 \text{ and } \frac{s_2^2}{n_2} = \frac{(6.40)^2}{10} = 4.10$$

We can now plug these values into the above equation as follows:

$$d.f. = \frac{[(4.45) + (4.10)]^2}{\frac{(4.45)^2}{11} + \frac{(4.10)^2}{9}} = \frac{73.10}{1.80 + 1.87} = 19.92$$

Because the number of degrees of freedom must be an integer, we round this result to 20. The critical t-score, taken from Table 4 in Appendix B, for a one-tail (right) test using $\alpha = 0.01$ with $d.f. = 20$ is $+2.528$. Because $t > t_c$, we reject H_0.

Letting Excel Do the Grunt Work

Excel performs many of the hypothesis tests that are discussed in this chapter. In this section, I'll demonstrate how to perform the previous battery example using this nifty tool. Follow these steps:

1. Open a blank Excel sheet and enter the data from the battery example in Columns A and B as shown in Figure 17.5.

2. From the Tools menu, choose Data Analysis and select t-Test: Two-Sample Assuming Unequal Variances. (Refer to the section "Installing the Data Analysis Add-in" from Chapter 2 if you don't see the Data Analysis command on the Tools menu.)

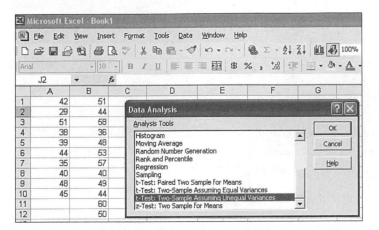

Figure 17.5

Data entry for the battery example.

3. Click OK.

4. In the t-Test: Two-Sample Assuming Unequal Variances dialog box, choose cells B1:B12 for Variable 1 Range and cells A1:A10 for Variable 2 Range. Set the Hypothesized Mean Difference to 0, Alpha to 0.01, and Output Range to cell D1, as shown in Figure 17.6.

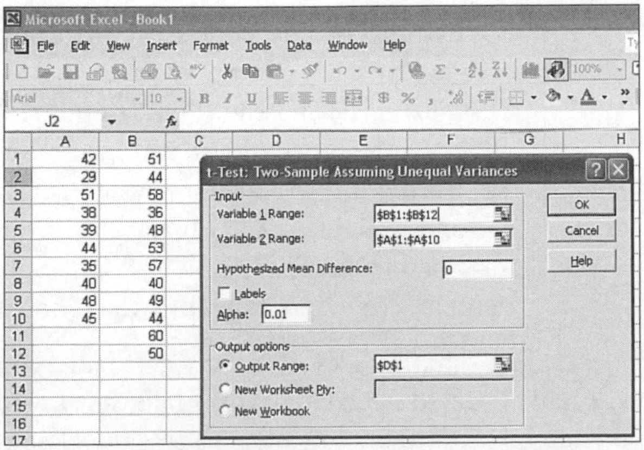

Figure 17.6

The t-test: Two-Sample Assuming Unequal Variances dialog box.

5. Click OK. The t-test output is shown in Figure 17.7.

Figure 17.7

t-test output.

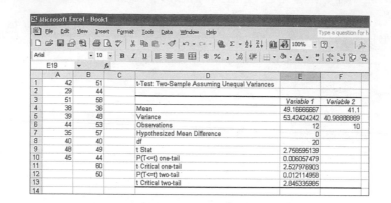

According to Figure 17.7, the calculated t-score of 2.758 is found in cell E9, which differs slightly from what we calculated in the previous section (2.77) due to the rounding of numbers. The *p*-value of 0.006 is found in cell E10. Because *p*-value ≤ α, we reject the null hypothesis.

Testing for Differences Between Means with Dependent Samples

Up to this point in the chapter, all the samples that we have used were *independent samples*. Samples are independent if they are not related in any way with each other. This is in contrast to *dependent samples*, where each observation of one sample is related to an observation in another.

Stat Facts

With **dependent samples,** the observation from one sample is related to an observation from another sample. With **independent samples,** there is no relationship in the observations between the samples.

An example of a dependent sample would be a weight-loss study. Each person is weighed at the beginning (Population 1) and end (Population 2) of the program. The change in weight of each person is calculated by subtracting the Population 2 weights from the Population 1 weights. Each observation from Population 1 is matched to an observation in Population 2. Dependent samples are tested differently than independent samples.

To demonstrate testing dependent samples, let's revisit my golf ball example from Chapter 15. If you remember, I had dreamed I invented a golf ball that I claimed would increase the distance off the tee by more than 20 yards. To test my claim, suppose we had 9 golfers hit my golf ball and the same golfers hit a regular golf ball. The following table shows these results. The letter d refers to the difference between my ball and the other ball.

Distance in Yards for Golf Ball Example

Golfer	1	2	3	4	5	6	7	8	9
My ball	215	228	256	264	248	255	239	218	239
Other ball	201	213	230	233	218	226	212	195	208
d	14	15	26	31	30	29	27	23	31
d^2	196	225	676	961	900	841	729	529	961

For future calculations, we will need:

$$\sum d = 14 + 15 + 26 + 31 + 30 + 29 + 27 + 23 + 31 = 226$$

$$\sum d^2 = 196 + 225 + 676 + 961 + 900 + 841 + 729 + 529 + 961 = 6018$$

The distances using my golf ball will be considered Population 1, and the distances with the other golf ball will be labeled Population 2. Because the same golfer hit both balls in each instance in the preceding table, these two samples are considered dependent.

My hypothesis statement for my claim would look like:

$$H_0 : \mu_1 - \mu_2 \leq 20$$

$$H_1 : \mu_1 - \mu_2 > 20$$

where:

μ_1 = the average distance off the tee with my new golf ball

μ_2 = the average distance off the tee with the other golf ball

However, because we are only interested in the difference between the two populations, we can rewrite this statement as a single sample hypothesis as follows:

$$H_0 : \mu_d \leq 20$$

$$H_1 : \mu_d > 20$$

where μ_d is the mean of the difference between the two populations.

We will test this hypothesis using $\alpha = 0.05$.

Our next step is to calculate the mean difference, $\bar{d}$, and the standard deviation of the difference, s_d, between the two samples as follows:

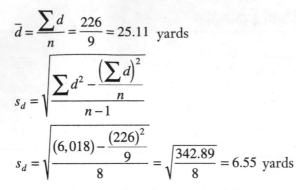

$$\bar{d} = \frac{\sum d}{n} = \frac{226}{9} = 25.11 \text{ yards}$$

$$s_d = \sqrt{\frac{\sum d^2 - \frac{\left(\sum d\right)^2}{n}}{n-1}}$$

$$s_d = \sqrt{\frac{(6,018) - \frac{(226)^2}{9}}{8}} = \sqrt{\frac{342.89}{8}} = 6.55 \text{ yards}$$

The equation for s_d is the same standard deviation equation that you learned in Chapter 5.

If both populations follow the normal distribution, we use the Student's t-distribution because both sample sizes are less than 30 and σ_1 and σ_2 are unknown. The calculated t-score is found using:

$$t = \frac{\bar{d} - \mu_d}{\frac{s_d}{\sqrt{n}}} = \frac{25.11 - 20}{\frac{6.55}{\sqrt{9}}} = \frac{5.11}{2.18} = +2.34$$

The number of degrees of freedom for this test is:

$$d.f. = n - 1 = 9 - 1 = 8$$

The critical t-score, taken from Table 4 in Appendix B, for a one-tail (right) test using $\alpha = 0.05$ with $d.f. = 8$ is +1.86. This hypothesis test is shown graphically in Figure 17.8.

Figure 17.8

Hypothesis test for the golf ball example.

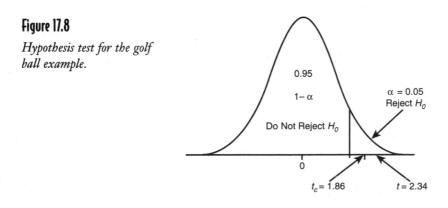

According to Figure 17.8, our calculated t-score of +2.34 is found in the "Reject H_0" region, which leads to our conclusion that my golf ball increases the distance off the tee by more than 20 yards. Too bad this was only a dream!

Testing for Differences Between Proportions with Independent Samples

Hypothesis testing can be performed to examine the difference between proportions of two populations as long as the sample size is large enough. Recall from Chapter 13, proportion data follow the binomial distribution, which can be approximated by the normal distribution under the following conditions:

$$np \geq 5 \text{ and } nq \geq 5$$

where:

p = the probability of a success in the population

q = the probability of a failure in the population ($q = 1 - p$)

Let's say that I want to test the claim that the proportion of males and females between the ages of 13 and 19 who use instant messages (IM) on the Internet every week are the same. My hypothesis would be stated as:

$$H_0 : p_1 = p_2$$
$$H_1 : p_1 \neq p_2$$

where:

p_1 = the proportion of 13- to 19-year-old males who use IMs every week

p_2 = the proportion of 13- to 19-year-old females who use IMs every week

The following table summarizes the data from the IM samples:

Summarized Data for IM Samples

Population	Number of Successes x	Sample Size n
Male	207	300
Female	266	350

What can we conclude at the $\alpha = 0.10$ level?

Our sample proportion of male IM users, $\bar{p}_1$, and female users, $\bar{p}_2$, can be found by:

$$\bar{p}_1 = \frac{x_1}{n_1} = \frac{207}{300} = 0.69 \text{ and } \bar{p}_2 = \frac{x_2}{n_2} = \frac{266}{350} = 0.76$$

To determine the calculated z-score, we need to know the *standard error of the difference between two proportions* (that's a mouthful), $\sigma_{\bar{p}_1 - \bar{p}_2}$, which is found using:

$$\sigma_{\bar{p}_1 - \bar{p}_2} = \sqrt{\frac{p_1(1-p_1)}{n_1} + \frac{p_2(1-p_2)}{n_2}}$$

Our problem is that we don't know the values of p_1 and p_2, the actual population proportions of male and female IM users. The next best thing is to calculate the *estimated standard error of the difference between two proportions*, $\sigma^{\mu}_{\bar{p}_1 - \bar{p}_2}$, using the following equation:

$$\sigma^{\mu}_{\bar{p}_1 - \bar{p}_2} = \sqrt{\left(\rho^{\mu}\right)\left(1 - \rho^{\mu}\right)\left(\frac{1}{n_1} + \frac{1}{n_2}\right)}$$

where ρ^{μ}, the *estimated overall proportion of two populations* is found using the following equation:

$$\rho^{\mu} = \frac{x_1 + x_2}{n_1 + n_2} = \frac{207 + 266}{300 + 350} = 0.728$$

For our IM example, the estimated standard error of the difference between two proportions is:

$$\sigma^{\mu}_{\bar{p}_1 - \bar{p}_2} = \sqrt{(0.728)(1 - 0.728)\left(\frac{1}{300} + \frac{1}{350}\right)} = 0.035$$

Bob's Basics

The term $\left(p_1 - p_2\right)_{H_0}$ refers to the hypothesized difference between the two population proportions. When the null hypothesis is testing that there is no difference between population proportions, then the term $\left(p_1 - p_2\right)_{H_0}$ is set to 0.

Now we can finally determine the calculated z-score using:

$$z = \frac{\left(\bar{p}_1 - \bar{p}_2\right) - \left(p_1 - p_2\right)_{H_0}}{\sigma^{\mu}_{\bar{p}_1 - \bar{p}_2}}$$

For the IM example, our calculated z-score becomes:

$$z = \frac{\left(\bar{p}^1 - \bar{p}^2\right) - \left(p^1 - p^2\right)_{H_0}}{\sigma^{\mu}_{\bar{p}_1 - \bar{p}_2}} = \frac{(0.69 - 0.76) - 0}{0.035} = -2.00$$

The critical z-score for a two-tail test with $\alpha = 0.10$ is ±1.64. This hypothesis test is shown graphically in Figure 17.9.

As you can see in Figure 17.9, the calculated z-score of –2.00 is within the "Reject H_0" region. There, we conclude that the proportions of male and female IM users between 13 and 19 years old are not equal to each other.

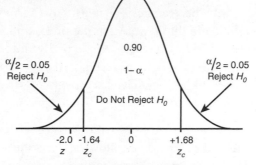

Figure 17.9

Hypothesis test for the IM example.

Stat Facts

The **standard error of the difference between two proportions** describes the variation in the difference between 2 sample proportions and is calculated using

$$\sigma_{\bar{p}_1 - \bar{p}_2} = \sqrt{\frac{p_1(1-p_1)}{n_1} + \frac{p_2(1-p_2)}{n_2}}$$. The **estimated standard error of the difference**

between two proportions approximates the variation in the difference between 2 sample

proportions and is calculated using $\sigma^{\mu}_{\bar{p}_1 - \bar{p}_2} = \sqrt{\left(\rho^{\mu}\right)\left(1 - \rho^{\mu}\right)\left(\frac{1}{n_1} + \frac{1}{n_2}\right)}$. The **estimated**

overall proportion of two populations is the weighted average of 2 sample proportions

and is calculated using $\rho^{\mu} = \dfrac{x_1 + x_2}{n_1 + n_2}$.

The *p*-value for these samples can be found using the normal *z*-score table found in Appendix B as follows:

$$2(P[z > +2.00]) = 2(1 - P[z \le +2.00])$$

$$2(P[z > +2.00]) = 2(1 - 0.9772) = 0.0456$$

This also confirms that we reject H_0 because the *p*-value $\le \alpha$.

This completes our invigorating journey through the land of hypothesis testing. Don't be too sad, though. We'll have the pleasure of revisiting this technique in Part 4 of this book—Advanced Inferential Statistics. I just bet you can't wait.

Your Turn

1. Test the hypothesis that the average SAT math scores from students in Pennsylvania and Ohio are different. A sample of 45 students from Pennsylvania had an average score of 552, whereas a sample of 38 Ohio students had an average score of 530. Assume the population standard deviations for Pennsylvania and Ohio are 105 and 114 respectively. Test at the $\alpha = 0.05$ level. What is the p-value for these samples?

2. A company tracks satisfaction scores based on customer feedback from individual stores on a scale of 0 to 100. The following data represents the customer scores from Store 1 and 2.

   Store 1:

   90   87   93   75   88   96   90   82   95   97   78

   Store 2:

   82   85   90   74   80   89   75   81   93   75

   Assume population standard deviations are equal but unknown and that the population is normally distributed. Test the hypothesis using $\alpha = 0.10$.

3. A new diet program claims that participants will lose more than 15 pounds after completion of the program. The following data represents the before and after weights of 9 individuals who completed the program. Test the claim at the $\alpha = 0.05$ level.

   Before   221 215 206 185 202 197 244 188 218

   After   200 192 195 166 187 177 227 165 201

4. Test the hypothesis that the proportion of home ownership in the state of Florida exceeds the national proportion at the $\alpha = 0.01$ level using the following data:

Population	Number of Successes	Sample Size
Florida	272	400
Nation	390	600

   What is the p-value for these samples?

The Least You Need to Know

- The normal distribution is used for the hypothesis test for the difference between means when $n \geq 30$ for both samples.

- The normal distribution is used for the hypothesis test for the difference between means when $n < 30$ for either sample, if σ_1 and σ_2 are known, and both populations are normally distributed.

- The Student's t-distribution is used for the hypothesis test for the difference between means when $n < 30$ for either sample, σ_1 and σ_2 are unknown, and both populations are normally distributed.

- With dependent samples, the observation from one sample is related to an observation from another sample. With independent samples, there is no relationship in the observations between the samples.

Part 4

Advanced Inferential Statistics

We covered a lot of ground so far in the first three parts of this book. What could possibly be left? Well, the last few topics focus on the more advanced statistical methods (don't worry, you can handle it) of chi-square tests, analysis of variance, and simple regression. Armed with these techniques, we can determine whether two categorical variables are related (chi-square), compare three or more populations (analysis of variance), and describe the strength and direction of the relationship between two variables (simple regression). After you have mastered these concepts, the sky is the limit!

Chapter

18

The Chi-Square Probability Distribution

In This Chapter

◆ Performing a goodness-of-fit test with the chi-square distribution

◆ Performing a test of independence with the chi-square distribution

◆ Using contingency tables to display frequency distributions

The last three chapters have explored the wonderful world of hypothesis testing. In these chapters, we compared means and proportions of one and two populations, making an educated conclusion about our initial claims. With that technique under our belt, we are now ready for bigger and better things.

In this chapter, we will compare two or more proportions using a new probability distribution: the chi-square. With this new test, we can confirm whether a set of data follows a specific probability distribution, such as the binomial or Poisson. (Remember those? They're back!) We can also use this distribution to determine whether two variables are statistically independent. It's actually a lot of fun—really it is!

Review of Data Measurement Scales

In Chapter 2, we discussed the different type of data measurement scales, which were nominal, ordinal, interval, and ratio. The following is a brief refresher of each:

◆ Nominal level of measurement deals strictly with qualitative data. Observations are simply assigned to predetermined categories. One example is gender of the respondent with the categories being male and female.

◆ Ordinal measurement is the next level up. It has all the properties of nominal data with the added feature that we can rank order the values from highest to lowest. An example would be ranking a movie as great, good, fair, or poor.

◆ Interval level of measurement involves strictly quantitative data. Here we can use the mathematical operations of addition and subtraction when comparing values. For this data, the difference between the different categories can be measured with actual numbers and also provides meaningful information. Temperature measurement in degrees Fahrenheit is a common example here.

◆ Ratio level is the highest measurement scale. Now we can perform all 4 mathematical operations to compare values. Examples of this type of data are age, weight, height, and salary. Ratio data has all the features of interval data with the added benefit of a "true zero point," meaning that a zero data value indicates the absence of the object being measured.

The hypothesis testing that we covered in the last three chapters strictly used interval and ratio data. However, the *chi-square distribution* in this chapter will allow us to perform hypothesis testing on nominal and ordinal data.

Stat Facts

The **chi-square distribution** is used to perform hypothesis testing on nominal and ordinal data.

The two major techniques that we will learn about in the chapter are using the chi-square distribution to perform a goodness-of-fit test and to test for the independence of two variables. So let's get started!

The Chi-Square Goodness-of-Fit Test

One of the many uses of the chi-square distribution is to perform a *goodness-of-fit test*, which uses a sample to test whether a frequency distribution fits the predicted distribution. As an example, let's say that a new movie in the making has an expected distribution of ratings summarized in the following table.

Expected Movie-Rating Distribution

Number of Stars	Percentage
5	40%
4	30%
3	20%
2	5%
1	5%
Total	100%

After its debut, a sample of 400 moviegoers were asked to rate the movie, with the results shown in the following table.

Observed Movie-Rating Distribution

Number of Stars	Number of Observations
5	145
4	128
3	73
2	32
1	22
Total	400

Can we conclude that the expected movie ratings are true based on the observed ratings of 400 people?

Stating the Null and Alternative Hypothesis

The null hypothesis in a chi-square goodness-of-fit test states that the sample of observed frequencies supports the claim about the expected frequencies. The alternative hypothesis states that there is no support for the claim pertaining to the expected frequencies. For our movie example, the hypothesis statement would look like the following:

Stat Facts

The **goodness-of-fit test** uses a sample to test whether a frequency distribution fits the predicted distribution.

H_0: The actual rating distribution can be described by the expected distribution.

H_1: The actual rating distribution differs from the expected distribution.

We will test this hypothesis at the $\alpha = 0.10$ level.

Observed Versus Expected Frequencies

Bob's Basics _____

The total number of expected (E) frequencies must be equal to the total number of observed (O) frequencies.

The chi-square test basically compares the observed (O) and expected (E) frequencies to determine whether there is a statistically significant difference. For our movie example, the *observed frequencies* are simply the number of observations collected for each category of our sample. The *expected frequencies* are the expected number of observations for each category and are calculated in the following table.

Stat Facts _____

Observed frequencies are the number of actual observations noted for each category of a frequency distribution with chi-squared analysis. **Expected frequencies** are the number of observations that would be expected for each category of a frequency distribution assuming the null hypothesis is true with chi-squared analysis.

Expected Frequency Table

Movie Rating	Expected Percentage	Sample Size	Expected Frequency (E)	Observed Frequency (O)
5	40%	400	0.40(400) = 160	145
4	30%	400	0.30(400) = 120	128
3	20%	400	0.20(400) = 80	73
2	5%	400	0.05(400) = 20	32
1	5%	400	0.05(400) = 20	22
Total	100%	400	400	

We are now ready to calculate the chi-square statistic.

Calculating the Chi-Square Statistic

The chi-square statistic is found using the following equation:

$$\chi^2 = \sum \frac{(O-E)^2}{E}$$

where:

O = the number of observed frequencies for each category

E = the number of expected frequencies for each category

The calculation using this equation is shown in the following table.

The Calculated Chi-Square Score for the Movie Example

Movie Rating	O	E	(O – E)	(O – E)²	$\frac{(O-E)^2}{E}$
5	145	160	–15	225	1.41
4	128	120	8	64	0.53
3	73	80	–7	49	0.61
2	32	20	12	144	7.20
1	22	20	2	4	0.20
Total					$\chi^2 = \sum \frac{(O-E)^2}{E} = 9.95$

Determining the Critical Chi-Square Score

The critical chi-square score, χ_c^2, depends on the number of degrees of freedom, which for this test would be:

$d.f. = k - 1$

where k equals the number of categories in the frequency distribution. For the movie example, there are 5 categories, so $d.f. = k - 1 = 5 - 1 = 4$.

The critical chi-square score is read from the chi-square table found on Table 5 in Appendix B of this book. An excerpt of this table is found here.

Critical Chi-Square Values

df	Selected right tail areas								
	0.3000	0.2000	0.1500	0.1000	0.0500	0.0250	0.0100	0.0050	0.0010
1	1.074	1.642	2.072	2.706	3.841	5.024	6.635	7.879	10.828
2	2.408	3.219	3.794	4.605	5.991	7.378	9.210	10.597	13.816
3	3.665	4.642	5.317	6.251	7.815	9.348	11.345	12.838	16.266
4	4.878	5.989	6.745	7.779	9.488	11.143	13.277	14.860	18.467
5	6.064	7.289	8.115	9.236	11.070	12.833	15.086	16.750	20.515
6	7.231	8.558	9.446	10.645	12.592	14.449	16.812	18.548	22.458

For $\alpha = 0.10$ and $d.f. = 4$, the critical chi-square score, $\chi_c^2 = 7.779$, as indicated in the underlined part of the table. Figure 18.1 shows the results of our hypothesis test.

Figure 18.1

Chi-square test for the movie example.

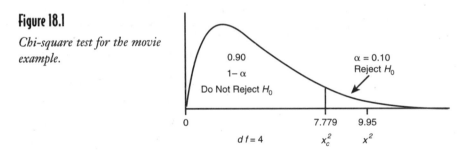

According to Figure 18.1, the calculated chi-square score of 9.95 is within the "Reject H_0" region, which leads us to the conclusion that the actual movie-rating frequency distribution differs than the expected distribution. We will always reject H_0 as long as $\chi_c^2 \leq \chi^2$.

Also, because the calculated chi-square score for the goodness-of-fit test can only be positive, the hypothesis test will always be a one-tail with the rejection region on the right side.

Using Excel's CHIINV Function

Don't have a chi-square distribution table handy? No need to panic. We can generate critical chi-square scores using Excel's CHIINV function, which has the following characteristics:

CHIINV(probability, deg_freedom)

where:

probability = the level of significance, α

deg_freedom = the number of degrees of freedom

For instance, Figure 18.2 shows the CHIINV function being used to determine the critical chi-square score for $\alpha = 0.10$ and *d.f.* = 4 from our previous example.

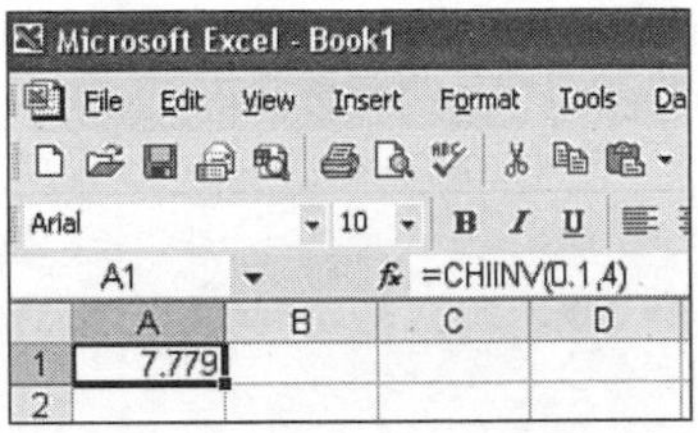

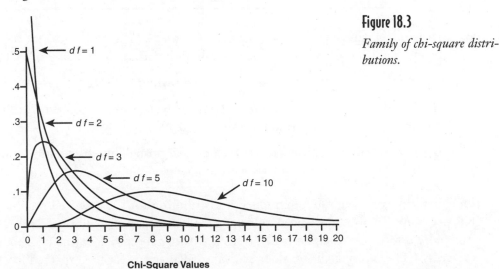

Figure 18.2

Excel's CHIINV function.

Cell A1 contains the Excel formula =CHIINV(0.10, 4) with the result being 7.779. This probability is underlined in the previous table.

Characteristics of a Chi-Square Distribution

We can see from Figure 18.2 that the chi-square distribution is not symmetrical but rather has a positive skew. The shape of the distribution will change with the number of degrees of freedom as shown in Figure 18.3.

Figure 18.3

Family of chi-square distributions.

As the number of degrees of freedom increases, the shape of the chi-square distribution becomes more symmetrical.

A Goodness-of-Fit Test with the Binomial Distribution

In past chapters, we have occasionally made assumptions that a population follows a specific distribution such as the normal or binomial. In this section, we can demonstrate how to verify this claim.

As an example, suppose that a certain Major League Baseball player claims the probability that he will get a hit at any given time is 30 percent. The following table is a frequency distribution of the number of hits per game over the last 100 games. Assume he has come to bat 4 times in each of the games.

Data for the Baseball Player

Number of Hits	Number of Games
0	26
1	34
2	30
3	7
4	3
Total	100

In other words, in 26 games he had 0 hits, in 34 games he had 1 hit, etc. Test the claim that this distribution follows a binomial distribution with $p = 0.30$ using $\alpha = 0.05$.

The hypothesis statement would look like the following:

H_0: The distribution of hits by the baseball player can be described with the binomial probability distribution using $p = 0.30$.

H_1: The distribution differs from the binomial probability distribution using $p = 0.30$.

Our first step is to calculate the frequency distribution for the expected number of hits per game. To do this, we need to look up the binomial probabilities in Table 1 from Appendix B for $n = 4$ (the number of trials per game) and $p = 0.30$ (the probability of a success). These probabilities, along with the calculations for the expected frequencies, are shown in the following table.

Expected Frequency Calculations for Baseball Player

Number of Hits per Game	Binomial Probabilities	Number of Games	Expected Frequency
0	0.2401 x	100 =	24.01
1	0.4116 x	100 =	41.16
2	0.2646 x	100 =	26.46
3	0.0756 x	100 =	7.56
4	0.0081 x	100 =	0.81
Total	1.0000		100.00

Before continuing, we need to make one adjustment to the expected frequencies. When using the chi-square test, we need at least 5 observations in each of the expected frequency categories. If there are less than 5, we need to combine categories. In the previous table, we will combine 3 and 4 hits per game into 1 category to meet this requirement.

Bob's Basics

Expected frequencies do not have to be integer numbers because they only represent theoretical values.

Now we are ready to determine the calculated chi-square score using the following table:

The Calculated Chi-Square Score for the Baseball Example

Hits	O	E	$(O - E)$	$(O - E)^2$	$\dfrac{(O-E)^2}{E}$
0	26	24.01	1.99	3.96	0.16
1	34	41.16	−7.16	51.27	1.25
2	30	26.46	3.54	12.53	0.47
3–4	10*	8.37**	1.63	2.66	0.32
Total					$\chi^2 = \sum \dfrac{(O-E)^2}{E} = 2.20$

*7 + 3 = 10
**7.56 + 0.81 = 8.37

According to Table 5 in Appendix B, the critical chi-square score for $\alpha = 0.05$ and $d.f. = k = 1 = 4 - 1 = 3$ is 7.815. This test is shown in Figure 18.4.

Figure 18.4

Chi-square test for the base-ball example.

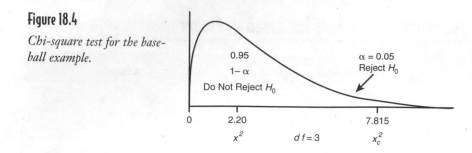

According to Figure 18.4, the calculated chi-square score of 2.20 is within the "Do Not Reject H_0" region, which leads us to the conclusion that the baseball player's hitting distribution can be described with the binomial distribution using $p = 0.30$.

Chi-Square Test for Independence

In addition to the goodness-of-fit test, the chi-square distribution can also test for independence between variables. To demonstrate this technique, I'm going to revisit the tennis example from Chapter 7.

If you recall, Deb felt that a short warm-up period before playing our match was hurting her chances of beating me. After examining the conditional probabilities, I had to admit there was some evidence supporting Deb's claim. However, I'm not one to take this sitting down. I demand justice, I demand further evidence, I demand a recount. (Oh wait a minute, this isn't Florida.) I demand … a hypothesis test using the chi-square distribution!

Unbeknownst to Deb, I have meticulously collected data from our 50 previous matches. The following table represents the number of wins for each of us according to the length of the warm-up period.

Observed Frequencies for Tennis Example

	0–10 Min	11–20 Min	More than 20 Min	Total
Deb wins	4	10	9	23
Bob wins	14	9	4	27
Total	18	19	13	50

This is known as a *contingency table*, which shows the observed frequencies of two variables. In this case, the variables are warm-up time and tennis player. The table is organized into *r* rows and *c* columns. For our table, *r* = 2 and *c* = 3. An intersection of a row and column is known as a *cell*. A contingency table has $r \times c$ cells, which in our case, would be 6.

Stat Facts

A contingency table shows the observed frequencies of two variables. An intersection of a row and column in a contingency table is known as a **cell**. A contingency table has r × c cells.

The chi-square test of independence will determine whether the proportion of times that Deb wins is the same for all 3 warm-up periods. If the outcome of the hypothesis test is that the proportions are not the same, we conclude that the length of warm-up does impact the performance of the players. But I have my doubts.

First we state the hypotheses as:

H_0: Warm-up time is independent of performance

H_1: Warm-up time affects performance

We will test this hypothesis at $\alpha = 0.10$ level.

Our next step is to determine the expected frequency of each cell in the contingency table under the assumption that the two variables are independent. We do this using the following equation:

$$E_{r,c} = \frac{\text{Total of Row } r \times \text{Total of Column } c}{\text{Total Number of Observations}}$$

where $E_{r,c}$ = the expected frequency of the cell that corresponds to the intersection of Row *r* and Column *c*.

The following table applies this notation to our tennis example.

Row/Column	Category	Total Observations
$r = 1$	Deb wins	23
$r = 2$	Bob wins	27
$c = 1$	0–10-minute warm-up	18
$c = 2$	11–20-minute warm-up	19
$c = 3$	More than 20-minute warm-up	13

The total number of observations for this example is 50, which we can confirm by adding 23 + 27 or 18 + 19 + 13. We can now determine the expected frequencies for each cell:

$$E_{1,1} = \frac{(23)(18)}{50} = 8.28 \quad E_{1,2} = \frac{(23)(19)}{50} = 8.74 \quad E_{1,3} = \frac{(23)(13)}{50} = 5.98$$

$$E_{2,1} = \frac{(27)(18)}{50} = 9.72 \quad E_{2,2} = \frac{(27)(19)}{50} = 10.26 \quad E_{2,3} = \frac{(27)(13)}{50} = 7.02$$

The following table summarizes these findings.

Expected Frequencies for Tennis Example

	0–10 Min	11–20 Min	More Than 20 Min	Total
Deb wins	8.28	8.74	5.98	23
Bob wins	9.72	10.26	7.02	27
Total	18	19	13	50

Bob's Basics

Notice that the expected frequencies for a contingency table add up to the row and column totals from the observed frequencies.

We now need to determine the calculated chi-square score using:

$$\chi^2 = \sum \frac{(O-E)^2}{E}$$

This calculation is summarized in the following table.

Chi-Square Calculation for the Tennis Example

Row	Column	O	E	$(O-E)$	$(O-E)^2$	$\frac{(O-E)^2}{E}$
1	1	4	8.28	−4.28	18.32	2.21
1	2	10	8.74	1.26	1.59	0.18
1	3	9	5.98	3.02	9.12	1.53
2	1	14	9.72	4.28	18.32	1.88
2	2	9	10.26	−1.26	1.59	0.15
2	3	4	7.02	−3.02	9.12	1.30

$$\chi^2 = \sum \frac{(O-E)^2}{E} = 7.25$$

To determine the critical chi-square score, we need to know the number of degrees of freedom, which for the independence test would be:

$$d.f. = (r - 1)(c - 1)$$

For this example, we have $(r - 1)(c - 1) = (2 - 1)(3 - 1) = 2$ degrees of freedom.

According to Table 5 in Appendix B, the critical chi-square score for $\alpha = 0.10$ and $d.f. = 2$ is 4.605. This test is shown in Figure 18.5.

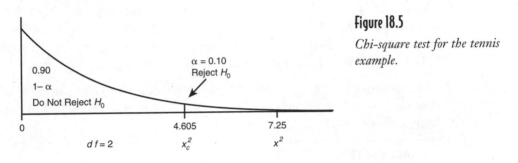

Figure 18.5

Chi-square test for the tennis example.

According to Figure 18.5, the calculated chi-square score of 7.25 is within the "Reject H_0" region, which leads us to the conclusion that there is a relationship between warm-up time and performance when Deb and I play tennis. Darn it—once again, Deb is right. Boy, does that have a familiar ring to it.

However, I do have one consolation. The chi-square test of independence only investigates whether a relationship exists between two variables. It does *not* conclude anything about the direction of the relationship. In other words, from a statistical perspective, Deb cannot claim that she is disadvantaged by the short warm-up time. She can only claim that warm-up time has some effect on her performance. We statisticians always leave ourselves a way out!

Your Turn

1. A company believes that the distribution of customer arrivals during the week are as follows:

Day	Expected Percentage of Customers
Monday	10
Tuesday	10
Wednesday	15

continues

continued

Day	Expected Percentage of Customers
Thursday	15
Friday	20
Saturday	30
Total	100

A week was randomly chosen and the number of customers each day was counted. The results were: Monday—31, Tuesday—18, Wednesday—36, Thursday—23, Friday—47, Saturday—60. Use this sample to test the expected distribution using $\alpha = 0.05$.

2. An e-commerce site would like to test the hypothesis that the number of hits per minute on their site follows the Poisson distribution with $\lambda = 3$. The following data was collected:

Number of Hits Per Minute	0	1	2	3	4	5	6	7 or More
Frequency	22	51	72	92	60	44	25	14

Test the hypothesis using $\alpha = 0.01$.

3. An English professor would like to test the relationship between an English grade and the number of hours per week a student reads. A survey of 500 students resulted in the following frequency distribution.

Numbers of Hours Reading per Week	Grade					Total
	A	B	C	D	F	
Less than 2	36	75	81	63	10	265
2–4	27	28	50	25	10	140
More than 4	32	25	24	6	8	95
Total	95	128	155	94	28	500

Test the hypothesis using $\alpha = 0.05$.

The Least You Need to Know

◆ The chi-square distribution is not symmetrical but rather has a positive skew. As the number of degrees of freedom increases, the shape of the chi-square distribution becomes more symmetrical.

◆ The chi-square distribution allows us to perform hypothesis testing on nominal and ordinal data.

◆ The chi-square distribution can be used to perform a goodness-of-fit test, which uses a sample to test whether a frequency distribution fits a predicted distribution.

◆ The chi-square test for independence investigates whether a relationship exists between two variables. It does not, however, test the direction of that relationship.

◆ A contingency table shows the observed frequencies of two variables. An intersection of a row and column in a contingency table is known as a cell.

Chapter 19

Analysis of Variance

In This Chapter

- ◆ Comparing three or more population means using analysis of variance
- ◆ Using the F-distribution to perform a hypothesis test for ANOVA
- ◆ Using Excel to perform a one-way ANOVA test
- ◆ Comparing pairs of sample means using the Scheffé test

In Chapter 17, we learned about a hypothesis test where we could compare the means of two different populations to see whether they were different. But what if we want to compare the means of three or more populations? Well you've come to the right place because that's what this chapter is all about.

To perform this new hypothesis test, we need to introduce one last probability distribution, known as the F-distribution. The test that we will perform has a very impressive name associated with it—the analysis of variance. This test is so special, it even has its own acronym, ANOVA. Sounds like something from outer space ... keep reading to find out.

One-Way Analysis of Variance

If you want to compare the means for 3 or more populations, ANOVA is the test for you. Let's say I'm interested in determining whether there is a difference in consumer satisfaction ratings between three fast-food chains. I would collect a sample of satisfaction ratings from each chain and test to see whether there is a significant difference between the sample means. Suppose my data look like the following.

Population	Fast-Food Chain	Sample Mean Rating
1	McDoogles	7.8
2	Burger Queen	8.2
3	Windy's	8.3

My hypothesis statement would look like the following:

$H_0 : \mu_1 = \mu_2 = \mu_3$

H_1 : not all μ's are equal

Essentially, I'm testing to see whether the variations in customer ratings from the previous table are due to the fast-food chains or whether the variations are purely random. In other words, do customers perceive any differences in satisfaction between the 3 chains? If I reject the null hypothesis, however, my only conclusion is that a difference does exist. Analysis of variance does not allow me to compare population means to one another to determine which is greater. That task requires further analysis.

Bob's Basics

To use one-way ANOVA, the following conditions must be present:

- ◆ The populations of interest must be normally distributed.
- ◆ The samples must be independent of each other.
- ◆ Each population must have the same variance.

A *factor* in ANOVA describes the cause of the variation in the data. In the previous example, the factor would be the fast-food chain. This would be considered a *one-way ANOVA* because there is only one factor being considered. More complex types of ANOVA can examine multiple factors.

A *level* in ANOVA describes the number of categories within the factor of interest. For our example, we have 3 levels based on the 3 different fast-food chains being examined.

To demonstrate one-way ANOVA, I'll use the following example. I admit, much to Deb's chagrin, that I am clueless when it comes to lawn care. My motto is "if it's green, it's good." Deb, on the other hand, knows exactly what type of fertilizer to get and when to apply it during the year. I hate spreading this stuff on the lawn because it apparently makes the grass grow faster, which means I have to cut it more often.

To make matters worse, we have a neighbor, Bill, who puts my yard to shame. Mr. "Perfect Lawn" is out every weekend, meticulously manicuring his domain until it looks like the home field for the National Lawn Bowling Association. This gives Deb a serious case of "lawn envy." Bill even has one of those cute little carts that he pulls on the back of his tractor. I asked Deb if I could get one for my tractor, but she says, based on my "Lawn IQ," I would probably just injure myself.

Anyway, there are several different types of analysis of variance and covering them all could be a book unto itself. The remainder of this chapter will use my lawn-care topic to describe a basic ANOVA procedure.

Stat Facts

A **factor** in ANOVA describes the cause of the variation in the data. When only one factor is being considered, the procedure is known as **one-way ANOVA**. A **level** in ANOVA describes the number of categories within the factor of interest.

Completely Randomized ANOVA

The simplest type of ANOVA is known as *completely randomized one-way ANOVA*, which involves an independent random selection of observations for each level of one factor. Now that's a mouthful! To help explain this, let's say I'm interested in comparing the effectiveness of 3 lawn fertilizers. Suppose I select 18 random patches of my precious lawn and apply either Fertilizer 1, 2, or 3 to each of them. After a week, I mow the patches and weigh the grass clippings.

The factor in this example is fertilizer. There are 3 levels, representing the 3 types of fertilizer being tested. The table that follows indicates the weight of the clippings in pounds from each patch. The mean and variance of each level are also shown.

Stat Facts

The simplest type of ANOVA is known as **completely randomized one-way ANOVA**, which involves an independent random selection of observations for each level of one factor.

Data for Lawn Clippings

	Fertilizer 1	Fertilizer 2	Fertilizer 3
	10.2	11.6	8.1
	8.5	12.0	9.0
	8.4	9.2	10.7
	10.5	10.3	9.1
	9.0	9.9	10.5
	8.1	12.5	9.5
Mean	9.12	10.92	9.48
Variance	1.01	1.70	0.96

The data for each type of fertilizer will be referred to as a sample. From the previous table, we have 3 samples, each consisting of 6 observations. The hypotheses statement can be stated as:

$H_0: \mu_1 = \mu_2 = \mu_3$

$H_1:$ not all μ's are equal

where μ_1, μ_2, and μ_3 are the true population means for the pounds of grass clippings for each type of fertilizer.

Partitioning the Sum of Squares

The hypothesis test for ANOVA compares two types of variations from the samples. We first need to recognize that the total variation in the data from our samples can be divided, or as statisticians like to say, "partitioned," into two parts.

The first part is the variation within each sample, which is officially known as the sum of squares within (SSW) and can be found using the following equation:

$$SSW = \sum_{i=1}^{k} (n_i - 1)s_i^2$$

where k = the number of samples (or levels).

For the fertilizer example, $k = 3$ and:

$s_1^2 = 1.01 \qquad s_2^2 = 1.70 \qquad s_3^2 = 0.96$

$n_1 = 6 \qquad n_2 = 6 \qquad n_3 = 6$

The sum of squares within can now be calculated as:

$$SSW = (6 - 1)1.01 + (6 - 1)1.70 + (6 - 1)0.96 = 18.35$$

Some textbooks will also refer to this value as the error sum of squares (SSE).

The second partition is the variation among the samples, which is known as the sum of squares between (SSB) and can be found by:

$$SSB = \sum_{i=1}^{k} n_i \left(\overline{x}_i - \overline{\overline{x}} \right)^2$$

where $\overline{\overline{x}}$ is the grand mean or the average value of all the observations. For the fertilizer example:

$$\overline{x}_1 = 9.12 \qquad \overline{x}_2 = 10.92 \qquad \overline{x}_3 = 9.48$$

We find $\overline{\overline{x}}$, the grand mean, using:

$$\overline{\overline{x}} = \frac{\sum x}{N}$$

> **Random Thoughts**
>
> Some textbooks will also refer to this SSB value as the treatment sum of squares (SSTR).

where N = the total number of observations from all samples.

For the fertilizer example:

$$\overline{\overline{x}} = \frac{10.2 + 8.5 + 8.4 + 10.5 + \ldots + 10.7 + 9.1 + 10.5 + 9.5}{18} = 9.83$$

We can now calculate the sum of squares between:

$$SSB = \sum_{i=1}^{k} n_i \left(\overline{x}_i - \overline{\overline{x}} \right)^2$$

$$SSB = 6(9.12 - 9.83)^2 + 6(10.92 - 9.83)^2 + 6(9.48 - 9.83)^2 = 10.86$$

Finally, the total variation of all the observations is known as the total sum of squares (SST) and can be found by:

$$SST = SSW + SSB$$

> **Random Thoughts**
>
> ANOVA does *not* require that all the sample sizes are equal, as they are in the fertilizer example. See Problem 1 in the "Your Turn" section as an example of unequal sample sizes.

For our example, this becomes:

$$SST = 18.35 + 10.86 = 29.21$$

Note that we can determine the variance of the original 18 observations, s^2, by:

$$s^2 = \frac{SST}{N-1} = \frac{29.21}{18-1} = 1.72$$

This result can be confirmed using the variance equation that was discussed way back in Chapter 5 or by using Excel.

Determining the Calculated F-Statistic

To test the hypothesis for ANOVA, we need to compare the calculated test statistic to a critical test statistic using the F-distribution. The calculated F-statistic can be found using the equation:

$$F = \frac{MSB}{MSW}$$

where MSB is the *mean square between*, found by:

$$MSB = \frac{SSB}{k-1}$$

and MSW is the *mean square within*, found by:

$$MSW = \frac{SSW}{N-k}$$

Now, let's put these guys to work with our fertilizer example.

$$MSB = \frac{SSB}{k-1} = \frac{10.86}{3-1} = 5.43$$

$$MSW = \frac{SSW}{N-k} = \frac{18.35}{18-3} = 1.22$$

$$F = \frac{MSB}{MSW} = \frac{5.43}{1.22} = 4.45$$

If the variation *between* the samples (MSB) is much greater than the variation *within* the samples (MSW), we will tend to reject the null hypothesis and conclude that there is a difference between population means.

Bob's Basics

The **mean square between (MSB)** is a measure of variation between the sample means. The **mean square within (MSW)** is a measure of variation within each sample. A large MSB variation, relative to the MSW variation, indicates that the sample means are not very close to one another. This condition will result in a large value of F, the calculated F-statistic. The larger the value of F, the more likely it will exceed the critical F-statistic (to be determined shortly), leading us to conclude there is a difference between population means.

To complete our test for this hypothesis, we need to introduce the F-distribution in the following section.

Determining the Critical F-Statistic

The F-distribution is used to determine the critical F-statistic, which is compared to the calculated F-statistic for the ANOVA hypothesis test. The critical F-statistic, $F_{\alpha, k-1, N-k}$, depends on two different degrees of freedom, which are determined by:

$$v_1 = k - 1 \text{ and } v_2 = N - k$$

For our fertilizer example:

$$v_1 = 3 - 1 = 2 \text{ and } v_2 = 18 - 3 = 15$$

The critical F-statistic is read from the F-distribution table found in Table 6 in Appendix B of this book. An excerpt of this table is found here.

Table of Critical F-Statistics

$\alpha = 0.05$

v_1

v_2	1	2	3	4	5	6	7	8	9	10
1	161.448	199.500	215.707	224.583	230.162	233.986	236.768	238.882	240.543	241.882
2	18.513	19.000	19.164	19.247	19.296	19.330	19.353	19.371	19.385	19.396
3	10.128	9.552	9.277	9.117	9.013	8.941	8.887	8.845	8.812	8.786
4	7.709	6.944	6.591	6.388	6.256	6.163	6.094	6.041	5.999	5.964
5	6.608	5.786	5.409	5.192	5.050	4.950	4.876	4.818	4.772	4.735
6	5.987	5.143	4.757	4.534	4.387	4.284	4.207	4.147	4.099	4.060
7	5.591	4.737	4.347	4.120	3.972	3.866	3.787	3.726	3.677	3.637
8	5.318	4.459	4.066	3.838	3.687	3.581	3.500	3.438	3.388	3.347
9	5.117	4.256	3.863	3.633	3.482	3.374	3.293	3.230	3.179	3.137
10	4.965	4.103	3.708	3.478	3.326	3.217	3.135	3.072	3.020	2.978
11	4.844	3.982	3.587	3.357	3.204	3.095	3.012	2.948	2.896	2.854
12	4.747	3.885	3.490	3.259	3.106	2.996	2.913	2.849	2.796	2.753
13	4.667	3.806	3.411	3.179	3.025	2.915	2.832	2.767	2.714	2.671
14	4.600	3.739	3.344	3.112	2.958	2.848	2.764	2.699	2.646	2.602
15	4.543	3.682	3.287	3.056	2.901	2.790	2.707	2.641	2.588	2.544
16	4.494	3.634	3.239	3.007	2.852	2.741	2.657	2.591	2.538	2.494
17	4.451	3.592	3.197	2.965	2.810	2.699	2.614	2.548	2.494	2.450
18	4.414	3.555	3.160	2.928	2.773	2.661	2.577	2.510	2.456	2.412

Note that this table is based only on $\alpha = 0.05$. Other values of α will require a different table. For $v_1 = 2$ and $v_2 = 15$, the critical F-statistic, $F_{.05,2,15} = 3.682$, as indicated in the underlined part of the table. Figure 19.1 shows the results of our hypothesis test.

Figure 19.1

ANOVA test for the fertilizer example.

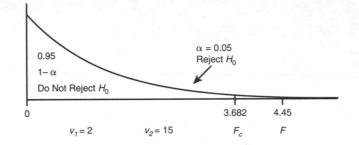

According to Figure 19.1, the calculated F-statistic of 4.45 is within the "Reject H_0" region, which leads us to the conclusion that the population means are not equal. We will always reject H_0 as long as $F_{\alpha, k-1, N-k} \leq F$.

Bob's Basics _____

The F-distribution has the following characteristics:

◆ It is not symmetrical but rather has a positive skew.

◆ The shape of the F-distribution will change with the degrees of freedom specified by the values of v_1 and v_2.

◆ As v_1 and v_2 increase in size, the shape of the F-distribution becomes more symmetrical.

◆ The total area under the curve is equal to 1.

◆ The F-distribution mean is approximately equal to 1.

Our final conclusion is that one or more of those darn fertilizers is making the grass grow faster than the others. Sounds like trouble to me.

Wrong Number _____

Even though we have rejected H_0 and concluded that the population means are not all equal, ANOVA does not allow us to make comparisons between means. In other words, we do not have enough evidence to conclude that Fertilizer 2 produces more grass clippings than Fertilizer 1. This requires another test, known as pairwise comparisons, which will be addressed later in this chapter.

Next we will explore how Excel can take some of the burden from all these nasty calculations.

Using Excel's FINV Function

Once again, we can generate critical F-statistics using Excel's FINV function, which has the following characteristics.

FINV(probability, deg_freedom1, deg_freedom2)

where:

probability = the level of significance,

deg_freedom1 = $v_1 = k - 1$

deg_freedom2 = $v_2 = N - k$

For instance, Figure 19.2 shows the FINV function being used to determine the critical F-statistic for $\alpha = 0.05$, $v_1 = 3 - 1 = 2$, and $v_2 = 18 - 3 = 15$ from our previous example.

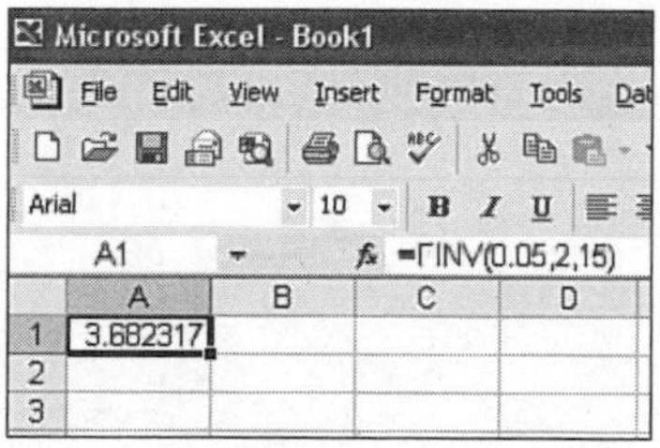

Figure 19.2

Excel's FINV function.

Cell A1 contains the Excel formula =FINV(0.05, 2, 15) with the result being 3.682. This probability is underlined in the previous table.

Using Excel to Perform One-Way ANOVA

I'm sure you've come to the conclusion that calculating ANOVA manually is a lot of work. I think you'll be amazed how easy this procedure is using Excel.

1. Start by placing the fertilizer data in Columns A, B, and C in a blank sheet.

2. Go to Tools menu and select Data Analysis. (Refer to the section "Installing the Data Analysis Add-in" from Chapter 2 if you don't see the Data Analysis command on the Tools menu.)

3. From the Data Analysis dialog box, select Anova: Single Factor as shown in Figure 19.3 and click OK.

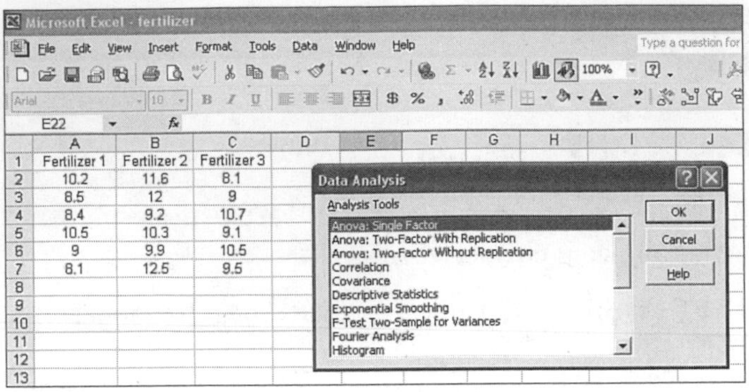

Figure 19.3

Setting up the one-way ANOVA in Excel.

4. Set up the Anova: Single Factor dialog box according to Figure 19.4.

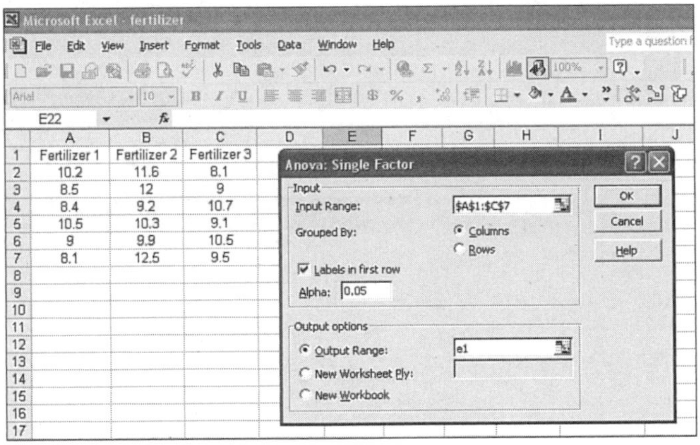

Figure 19.4

The AVOVA: Single Factor dialog box.

5. Click OK. Figure 19.5 shows the final ANOVA results.

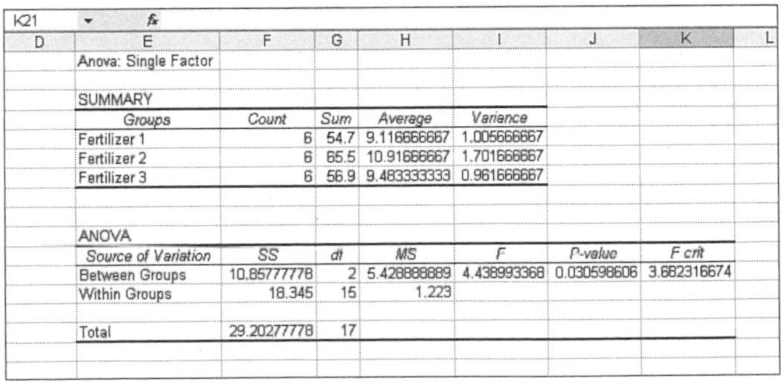

Anova: Single Factor

SUMMARY

Groups	Count	Sum	Average	Variance
Fertilizer 1	6	54.7	9.116666667	1.005666667
Fertilizer 2	6	65.5	10.91666667	1.701666667
Fertilizer 3	6	56.9	9.483333333	0.961666667

ANOVA

Source of Variation	SS	df	MS	F	P-value	F crit
Between Groups	10.85777778	2	5.428888889	4.438993368	0.030598606	3.682316674
Within Groups	18.345	15	1.223			
Total	29.20277778	17				

Figure 19.5

Final results of the one-way ANOVA in Excel.

These results are consistent with what we found doing it the hard way in the previous sections. Notice that the p-value = 0.0305 for this test, meaning we can reject H_0, because this p-value $\leq \alpha$. If you remember, we had set $\alpha = 0.05$ when we stated the hypothesis test.

Pairwise Comparisons

Once we have rejected H_0 using ANOVA, we can determine which of the sample means are different using the Scheffé test. This test compares each pair of sample means from the ANOVA procedure. For the fertilizer example, we would compare $\bar{x}_1$ versus $\bar{x}_2$, $\bar{x}_1$ versus $\bar{x}_3$, and $\bar{x}_2$ versus $\bar{x}_3$ to see whether any differences exist.

First, the following test statistic for the Scheffé test, F_s, is calculated for each of the pairs of sample means:

$$F_S = \frac{\left(\bar{x}_a - \bar{x}_b\right)^2}{\dfrac{SSW}{\sum\limits_{i=1}^{k}(n_i - 1)}\left[\dfrac{1}{n_a} + \dfrac{1}{n_b}\right]}$$

Bob's Basics

After rejecting H_0 using ANOVA, we can determine which of the sample means are different using the Scheffé test.

where:

$\bar{x}_a$, $\bar{x}_b$ = the sample means being compared

SSW = the sum of squares within from the ANOVA procedure

n_a, n_b = the samples sizes

k = the number of samples (or levels)

Comparing $\bar{x}_1$ and $\bar{x}_2$, we have:

$$F_S = \frac{\left(\bar{x}_a - \bar{x}_b\right)^2}{\dfrac{SSW}{\sum\limits_{i=1}^{k}(n_i - 1)}\left[\dfrac{1}{n_a} + \dfrac{1}{n_b}\right]} = \frac{(9.12 - 10.92)^2}{\dfrac{18.35}{5+5+5}\left[\dfrac{1}{6} + \dfrac{1}{6}\right]}$$

$$F_S = \frac{3.24}{1.22[0.33]} = 8.048$$

Comparing $\bar{x}_1$ and $\bar{x}_3$, we have:

$$F_S = \frac{(9.12 - 9.48)^2}{\dfrac{18.35}{5+5+5}\left[\dfrac{1}{6} + \dfrac{1}{6}\right]} = \frac{0.13}{1.22[0.33]} = 0.323$$

Comparing $\bar{x}_2$ and $\bar{x}_3$, we have:

$$F_S = \frac{(10.92 - 9.48)^2}{\frac{18.35}{5+5+5}\left[\frac{1}{6} + \frac{1}{6}\right]} = \frac{2.07}{1.22[0.33]} = 5.142$$

Next the critical value for the Scheffé test, F_{sc}, is determined by multiplying the critical F-statistic from the ANOVA test by k – 1 as follows:

$$F_{sc} = (k - 1)F_{\alpha, k-1, N-k}$$

For the fertilizer example:

$$F_{.05, 2, 15} = 3.682$$

$$F_{sc} = (3 - 1)(3.682) = 7.364$$

If $F_s \leq F_{sc}$, we conclude there is no difference between the sample means, otherwise there is a difference. The following table summarizes these results.

Summary of the Scheffé Test

Sample Pair	F_s	F_{sc}	Conclusion
$\bar{x}_1$ and $\bar{x}_2$	8.048	7.364	Difference
$\bar{x}_1$ and $\bar{x}_3$	0.323	7.364	No Difference
$\bar{x}_2$ and $\bar{x}_3$	5.142	7.364	No Difference

According to our results, the only statistically significant difference is between Fertilizer 1 and Fertilizer 2. It appears that Fertilizer 2 is more effective in making grass grow faster when compared to Fertilizer 1. I better keep Deb away from this brand.

Your Turn

1. A consumer group is testing the gas mileage of 3 different models of cars. Several cars of each model were driven 500 miles and the mileage was recorded as follows.

Car 1	Car 2	Car 3
22.5	18.7	17.2
20.8	19.8	18.0
22.0	20.4	21.1
23.6	18.0	19.8

Car 1	Car 2	Car 3
21.3	21.4	18.6
22.5	19.7	

Note that the size of each sample does not have to be equal for ANOVA.

Test for a difference between sample means using $\alpha = 0.05$.

2. Perform a pairwise comparison test on the sample means from Problem 1.

3. A vice president would like to determine whether there is a difference between the average number of customers per day between 4 different stores using the following data.

Store 1	Store 2	Store 3	Store 4
36	35	26	26
48	20	20	52
32	31	38	37
28	22	32	36
31	19	37	18
55	42	15	30
	29		21

Note that the size of each sample does not have to be equal for ANOVA.

Test for a difference between sample means using $\alpha = 0.05$.

The Least You Need to Know

- Analysis of variance, also known as ANOVA, compares the means of 3 or more populations.

- A factor in ANOVA describes the cause of the variation in the data. When only one factor is being considered, the procedure is known as one-way ANOVA.

- A level in ANOVA describes the number of categories within the factor of interest.

- The simplest type of ANOVA is known as completely randomized one-way ANOVA, which involves an independent random selection of observations for each level of one factor.

- To test the hypothesis for ANOVA, we need to compare the calculated test statistic to a critical test statistic using the F-distribution.

- After rejecting H_0 using ANOVA, we can determine which of the sample means are different using the Scheffé test.

20

Correlation and Simple Regression

In This Chapter

- ◆ Distinguishing between independent and dependent variables
- ◆ Determining the correlation and regression line for a set of ordered-pair data
- ◆ Calculating a confidence interval for a regression line
- ◆ Performing a hypothesis test on the regression line
- ◆ Using Excel to perform simple regression analysis

For the last several chapters, we have put inferential statistics to work drawing conclusions about one, two, or more population means and proportions. I know this has been a lot of fun for you, but it's time to move to another type of inferential statistics that's even more exciting. (If you can imagine that!)

This final chapter focuses on describing how two variables relate to one another. Using correlation and simple regression, we will be able to first, determine whether a relationship does indeed exist between the variables and second, describe the nature of this relationship in mathematical terms. And, hopefully, we'll have some fun doing it!

Independent Versus Dependent Variables

Suppose I would like to investigate the relationship between the number of hours that a student studies for a statistics exam and the grade for that exam (uh oh). The following table shows sample data from 6 students who were randomly chosen.

Data for Statistics Exam

Hours Studied	Exam Grade
3	86
5	95
4	92
4	83
2	78
3	82

Stat Facts

The **independent variable** (x) causes variation in the **dependent variable** (y).

Wrong Number

Exercise caution when deciding which variable is independent and which is dependent. Examine the relationship from both directions to see which one makes the most sense. The wrong choice will lead to meaningless results.

Obviously, we would expect the number of hours studying to affect the grade. The Hours Studied variable is considered the *independent variable* (x) because it causes the observed variation in the Exam Grade, which is considered to be the *dependent variable* (y). The data from the previous table is considered *ordered pairs* of (x,y) values, such as (3,86) and (5,95).

This "causal relationship" between independent and dependent variables only exists in one direction as shown here:

Independent variable (x) → Dependent variable (y)

This relationship does not work in reverse. For instance, we would not expect that the exam grade variable would cause the student to study a certain number of hours in our previous example.

Other examples of independent and dependent variables are shown in the following table.

Examples of Independent and Dependent Variables

Independent Variable	Dependent Variable
Size of TV	Selling price of TV
Level of advertising	Volume of sales
Size of sports team payroll	Number of games won

Our next section focuses on describing the relationship between the x and y variables using inferential statistics.

Correlation

Correlation measures both the strength and direction of the relationship between x and y. Figure 20.1 illustrates the different types of correlation in a series of scatter plots, which graphs each ordered pair of (x, y). The convention is to place the x variable on the horizontal axis and the y variable on the vertical axis.

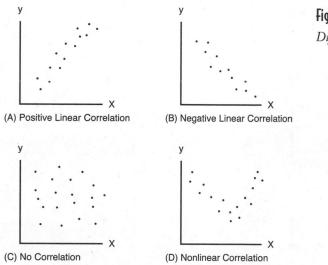

Figure 20.1

Different types of correlation.

Graph A in Figure 20.1 shows an example of positive linear correlation, where as x increases, y also tends to increase in a linear (straight line) fashion. Graph B shows a negative linear correlation, where as x increases, y tends to decrease linearly. Graph C indicates no correlation between x and y. This set of variables appears to have no impact on each other. And finally, Graph D is an example of a nonlinear relationship between variables. As x increases, y decreases at first, and then changes direction and increases.

The remainder of this chapter focuses on linear relationships between the independent and dependent variables. Nonlinear relationships can be very disagreeable and go beyond the scope of this book.

Correlation Coefficient

The correlation coefficient, r, provides us with both the strength and direction of the relationship between the independent and dependent variables. Values of r range between −1.0 and +1.0. When r is positive, the relationship between x and y is positive (Graph A from Figure 20.1), and when r is negative, the relationship is negative (Graph B). A correlation coefficient close to 0 is evidence that there is no relationship between x and y (Graph C).

The strength of the relationship between x and y is measured by how close the correlation coefficient is to +1.0 or −1.0 and can be viewed in Figure 20.2.

Figure 20.2

The strength of the relationship.

The Strength of the Relationship

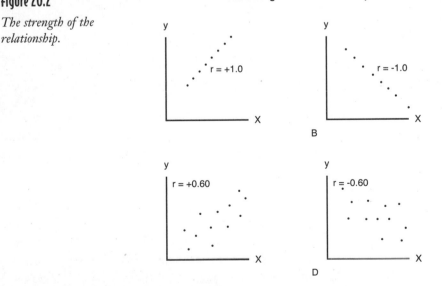

Graph A illustrates a perfect positive correlation between x and y with $r = +10$. Graph B shows a perfect negative correlation between x and y with $r = -10$. Graphs C and D are examples of weaker relationships between the independent and dependent variables.

Stat Facts

The correlation coefficient, r, indicates both the strength and direction of the relationship between the independent and dependent variables. Values of r range from -1.0, a strong negative relationship, to $+1.0$, a strong positive relationship. When $r = 0$, there is no relationship between variables x and y.

We can calculate the actual correlation coefficient using the following equation:

$$r = \frac{n\sum xy - \left(\sum x\right)\left(\sum y\right)}{\sqrt{\left[n\sum x^2 - \left(\sum x\right)^2\right]\left[n\sum y^2 - \left(\sum y\right)^2\right]}}$$

Wow. I know this looks overwhelming, but before we panic, let's try out our exam grade example on this. The following table will help break down the calculations and make them more manageable.

Hours of Study	Exam	Grade		
x	y	xy	x^2	y^2
3	86	258	9	7,396
5	95	368	25	8,464
4	92	475	16	9,025
4	83	332	16	6,889
2	78	156	4	6,084
3	82	246	9	6,724
$\sum x = 21$	$\sum y = 516$	$\sum xy = 1{,}835$	$\sum x^2 = 79$	$\sum y^2 = 44{,}582$

Using these values along with $n = 6$, the number of ordered pairs, we have:

$$r = \frac{n\sum xy - \left(\sum x\right)\left(\sum y\right)}{\sqrt{\left[n\sum x^2 - \left(\sum x\right)^2\right]\left[n\sum y^2 - \left(\sum y\right)^2\right]}}$$

$$r = \frac{6(1{,}835) - (21)(516)}{\sqrt{\left[6(79) - (21)^2\right]\left[6(44{,}582) - (516)^2\right]}}$$

$$r = \frac{174}{\sqrt{(33)(1{,}236)}} = 0.862$$

As you can see, we have a fairly strong positive correlation between hours of study and exam grade. That's good news for us teachers.

> **CAUTION**
>
> **Wrong Number** _____
>
> Be careful to distinguish between $\sum x^2$ and $\left(\sum x\right)^2$. With $\sum x^2$, we first square each value of x and then add each squared term. With $\left(\sum x\right)^2$, we first add each value of x, and then square this result. The answers between the two are very different!

What is the benefit of establishing a relationship between two variables such as these? Excellent question. When we discover that a relationship does exist, we can predict exam scores based on a particular number of hours of study. Simply put, the stronger the relationship, the more accurate our prediction will be. You will learn how to make such predictions later in this chapter when we discuss simple regression.

Testing the Significance of the Correlation Coefficient

We can perform a hypothesis test to determine whether the population correlation coefficient, ρ, is significantly different from 0 based on the value of the calculated correlation coefficient, r. The hypotheses can be stated as:

$H_0: \rho \le 0$

$H_1: \rho > 0$

This statement tests whether a positive correlation exists between x and y. I could also choose a two-tail test that would investigate whether any correlation exists (either positive or negative) by setting $H_0: \rho = 0$ and $H_1: \rho \ne 0$.

The test statistic for the correlation coefficient uses the Student's t-distribution as follows:

$$t = \frac{r}{\sqrt{\dfrac{1-r^2}{n-2}}}$$

where:

    r = the calculated correlation coefficient from the ordered pairs

    n = the number of ordered pairs

For the exam grade example, the calculated t-statistic becomes:

$$t = \frac{r}{\sqrt{\dfrac{1-r^2}{n-2}}} = \frac{0.862}{\sqrt{\dfrac{1-(0.862)^2}{6-2}}}$$

$$t = \frac{0.862}{\sqrt{\dfrac{.257}{4}}} = 3.401$$

The critical t-statistic is based on $d.f. = n - 2$. If we choose $\alpha = 0.05$, $t_c = 2.132$ from Table 4 in Appendix B for a one-tail test. Because $t > t_c$, we reject H_0 and conclude that there is indeed a positive correlation coefficient between hours of study and the exam grade. Once again, statistics has proven that all is right in the world!

Using Excel to Calculate Correlation Coefficients

After looking at the nasty calculations involved for the correlation coefficient, I'm sure you'll be relieved to know that Excel will do the work for you with the CORREL function that has the following characteristics:

CORREL(array1, array2)

where:

array1 = the range of data for the first variable

array2 = the range of data for the second variable

For instance, Figure 20.3 shows the CORREL function being used to calculate the correlation coefficient for the exam grade example.

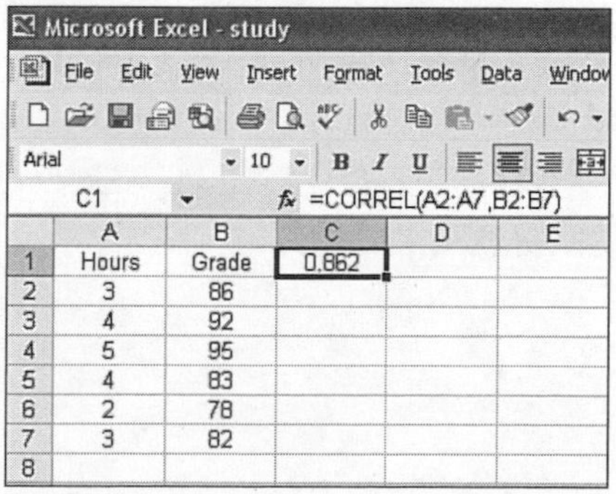

Figure 20.3

CORREL function in Excel with the exam grade example.

Cell C1 contains the Excel formula =CORREL(A2:A7,B2:B7) with the result being 0.862.

Simple Regression

The technique of *simple regression* enables us to describe a straight line that best fits a series of ordered pairs (x,y). The equation for a straight line, known as a *linear equation*, takes the form:

$$\hat{y} = a + bx$$

Stat Facts

The technique of **simple regression** enables us to describe a straight line that best fits a series of ordered pairs (x,y).

where:

$\hat{y}$ = the predicted value of y, given a value of x

x = the independent variable

a = the y-intercept for the straight line

b = the slope of the straight line

Figure 20.4 illustrates this concept.

Figure 20.4

Equation for a straight line.

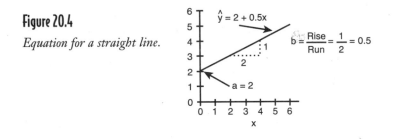

Figure 20.4 shows a line described by the equation $\hat{y} = 2 + 0.5x$. The y-intercept is the point where the line crosses the y-axis, which in this case is a = 2. The slope of the line, b, is shown as the ratio of the rise of the line over the run of the line, shown as b = 0.5. A positive slope indicates the line is rising from left to right. A negative slope, you guessed it, moves lower from left to right. If b = 0, the line is horizontal, which means there is no relationship between the independent and dependent variables. In other words, a change in the value of x has no effect on the value of y.

Students sometimes struggle with the distinction between $\hat{y}$ and y. Figure 20.5 shows 6 ordered pairs and a line that appear to fit the data described by the equation $\hat{y} = 2 + 0.5x$.

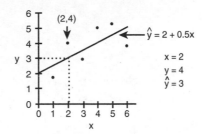

Figure 20.5

The difference between y *and* $\hat{y}$.

Figure 20.5 shows a data point that corresponds to the ordered pair $x = 2$ and $y = 4$. Notice that the *predicted* value of y according to the line at $x = 2$ is $\hat{y} = 3$. We can verify this using the equation as follows:

$$\hat{y} = 20 + 0.5x = 2 + 0.5(2) = 3$$

The value of y represents an actual data point while the value of $\hat{y}$ is the predicted value of y using the linear equation, given a value for x.

Our next step is to find the linear equation that best fits a set of ordered pairs.

The Least Squares Method

The *least squares method* is a mathematical procedure to identify the linear equation that best fits a set of ordered pairs by finding values for a, the y-intercept, and b, the slope. The goal of the least squares method is to minimize the total squared error between the values of y and $\hat{y}$. If we define the error as $y - \hat{y}$ for each data point, the least squares method will minimize:

$$\sum_{i=1}^{n}\left(y_i - \hat{y}_i\right)^2$$

where n is the number of ordered pairs around the line that best fits the data.

This concept is illustrated in Figure 20.6.

Stat Facts

The **least squares method** is a mathematical procedure to identify the linear equation that best fits a set of ordered pairs by finding values for a, the y-intercept, and b, the slope. The goal of the least squares method is to minimize the total squared error between the values of y and $\hat{y}$. The **regression line** is the line that best fits the data.

Figure 20.6

Minimizing the error.

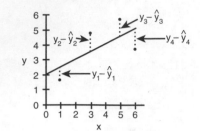

According to Figure 20.6, the line that best fits the data, the *regression line*, will minimize the total squared error of the 4 data points. I'll demonstrate how to determine this regression equation using the least squares method through the following example.

Apparently, there has been a silent war raging in our bathroom at home that has recently caught my attention. I'm of course referring to the space on our bathroom countertop that, under an unwritten agreement, Deb and I were supposed to "share." Over the past few months, I have been keeping a wary eye on the increasing number of "things" on her side that are growing in number at a rate faster than the federal deficit. I'm slowly being squeezed out of my end of the bathroom by containers with words such as "volumizing fixative" and "soyagen complex." I might as well just raise the white towel in surrender and head off to the teenagers' bathroom in exile, a room I had vowed *never* to step foot into because … well, I'll just spare you the messy details.

Anyway, let's say the following table shows the number of Deb's items on the bathroom counter for the past several months.

Bathroom Counter Data

Month	Number of Items	Month	Number of Items
1	8	6	13
2	6	7	9
3	10	8	11
4	6	9	15
5	10	10	17

Because my goal is to investigate whether the number of items is increasing over time, Month will be the independent variable and Number of Items will be the dependent variable.

The least squares method finds the linear equation that best fits the data by determining the value for a, the y-intercept, and b, the slope, using the following equations:

$$b = \frac{n\sum xy - \left(\sum x\right)\left(\sum y\right)}{n\sum x^2 - \left(\sum x\right)^2}$$

$$a = \bar{y} - b\bar{x}$$

where:

$\bar{x}$ = the average value of x, the independent variable

$\bar{y}$ = the average value of y, the dependent variable

The following table summarizes the calculations necessary for these equations.

Calculations for the Slope and Intercept

Month	Items			
x	y	xy	x^2	y^2
1	8	8	1	84
2	6	12	4	36
3	10	30	9	100
4	6	24	16	36
5	10	50	25	100
6	13	78	36	169
7	9	63	49	81
8	11	88	64	121
9	15	135	81	225
10	17	170	100	289
$\sum x = 55$	$\sum y = 105$	$\sum xy = 658$	$\sum x^2 = 385$	$\sum y^2 = 1,221$

$$\bar{x} = \frac{55}{10} = 5.5 \qquad\qquad \bar{y} = \frac{105}{10} = 10.5$$

$$b = \frac{n\sum xy - \left(\sum x\right)\left(\sum y\right)}{n\sum x^2 - \left(\sum x\right)^2} = \frac{10(658) - (55)(105)}{10(385) - (55)^2}$$

$$b = \frac{805}{825} = 0.976$$

$$a = \bar{y} - b\bar{x} = 10.5 - (0.976)5.5 = 5.13$$

The regression line for the bathroom counter example would be:

$$\hat{y} = 5.13 + 0.976x$$

Because the slope of this equation is a positive 0.976, I have evidence that the number of items on the counter is increasing over time at an average rate of nearly 1 per month. Figure 20.7 shows the regression line with the ordered pairs.

Figure 20.7

Regression line for the bath-room counter example.

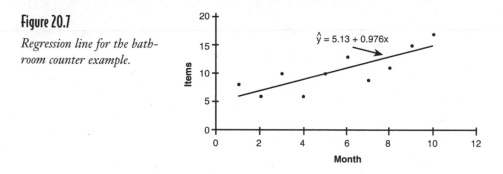

My prediction for the number of items on the counter in another 6 months (Month 16 from my data) will be:

$$\hat{y} = 5.13 + 0.976x = 5.13 + 0.976(16) = 20.7 \approx 21 \text{ items}$$

Look out kids. Make room for Dad.

Confidence Interval for the Regression Line

Just how accurate is my estimate for the number of items on the counter for a particular month? To answer this, we need to determine the *standard error of the estimate, s_e,* using the following formula:

$$s_e = \sqrt{\frac{\sum y^2 - a\sum y - b\sum xy}{n-2}}$$

The standard error of the estimate measures the amount of dispersion of the observed data around the regression line. If the data points are very close to the line, the standard error of the estimate is relatively low and vice versa. For our bathroom example:

$$s_e = \sqrt{\frac{\sum y^2 - a\sum y - b\sum xy}{n-2}} = \sqrt{\frac{(1221) - 5.13(105) - 0.976(658)}{10-2}}$$

$$s_e = \sqrt{\frac{40.14}{8}} = 2.24$$

We are now ready to calculate a confidence interval (Remember those from Chapter 14?) for the mean of y around a particular value of x. For Month 8 ($x = 8$) in the data, Deb has 11 items ($y = 11$) on the counter. The regression line predicted she would have:

$$\hat{y} = 5.13 + 0.976x = 5.13 + 0.976(8) = 12.9 \text{ items}$$

In general, the confidence interval around the mean of y given a specific value of x can be found by:

$$CI = \hat{y} \pm t_c s_e \sqrt{\frac{1}{n} + \frac{(x - \bar{x})^2}{(\sum x^2) - \frac{(\sum x)^2}{n}}}$$

Stat Facts

The **standard error of the estimate, s_e,** measures the amount of dispersion of the observed data around the regression line.

where:

t_c = the critical t-statistic from the Students' t-distribution

s_e = the standard error of the mean

n = the number of ordered pairs

Hold on to your hat while we dive into this one with our example. Suppose we would like a 95 percent confidence interval around the mean of y for Month 8. To find our critical t-statistic, we look to Table 4 in Appendix B. This procedure has $n - 2 = 10 - 2 = 8$ degrees of freedom, resulting in $t_c = 2.306$ from Table 4 in Appendix B. Our confidence interval is then:

$$CI = \hat{y} \pm t_c s_e \sqrt{\frac{1}{n} + \frac{(x - \bar{x})^2}{(\sum x^2) - \frac{(\sum x)^2}{n}}}$$

$$CI = 12.9 \pm (2.306)(2.24)\sqrt{\frac{1}{10} + \frac{(8 - 5.5)^2}{(385) - \frac{(55)^2}{10}}}$$

$$CI = 12.9 \pm (2.306)(2.24)(0.419) = 12.9 \pm 2.16$$

$$CI = 10.74 \text{ and } 15.06$$

This interval is shown graphically on Figure 20.8.

Figure 20.8

95 percent confidence interval for x = 8.

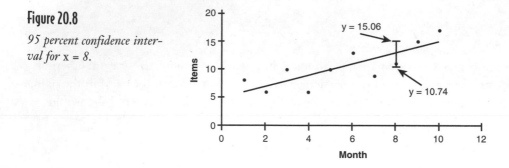

Our 95 percent confidence interval for the number of items on the counter in Month 8 is between 10.74 and 15.06 items. Sounds like a very crowded countertop to me.

Testing the Slope of the Regression Line

Recall that if the slope of the regression line, *b*, is equal to 0, then there is no relationship between *x* and *y*. In our bathroom counter example, we found the slope of the regression line to be 0.976. However, because this result was based on a sample of observations, we need to test to see whether 0.976 is far enough away from 0 to claim a relationship really does exist between the variables. If β is the slope of the true population, then our hypotheses statement would be:

$H_0: \beta = 0$

$H_1: \beta \neq 0$

If we reject the null hypothesis, we conclude that a relationship does exist between the independent and dependent variables based on our sample. We'll test this using $\alpha = 0.01$.

This hypothesis test requires the *standard error of the slope*, s_b, which is found with the following equation:

$$s_b = \frac{s_e}{\sqrt{\sum x^2 - n\bar{x}^2}}$$

where s_e is the standard error of the estimate that we calculated earlier.

For our bathroom example:

$$s_b = \frac{s_e}{\sqrt{\sum x^2 - n\bar{x}^2}} = \frac{2.24}{\sqrt{385 - 10(5.5)^2}}$$

$$s_b \frac{2.24}{\sqrt{82.5}} = 0.247$$

The test statistic for this hypothesis is:

$$t = \frac{b - \beta_{H_0}}{s_b}$$

where β_{H_0} is the value of the population slope according to the null hypothesis.

For this example, our calculated t-statistic is:

$$t = \frac{b - \beta_{H_0}}{s_b} = \frac{0.976 - 0}{0.247} = 3.951$$

The critical t-statistic is taken from the Students' t-distribution with $n - 2 = 10 - 2 = 8$ degrees of freedom. With a two-tail test and $\alpha = 0.10$, $t_c = 3.355$, according to Table 4 in Appendix B. Because $t > t_c$, we reject the null hypothesis and conclude there is a relationship between the month and the number of items on the bathroom countertop. I thought so!

> **Wrong Number**
>
> Just because a relationship between two variables is statistically significant doesn't necessarily mean that a causal relationship truly exists. The mathematical relationship could be due to pure coincidence. Always use your best judgment when making these decisions.

The Coefficient of Determination

Another way of measuring the strength of a relationship is with the *coefficient of determination*, r^2. This represents the percentage of the variation in y that is explained by the regression line. We find this value by simply squaring r, the correlation coefficient. For the bathroom example, the correlation coefficient is:

$$r = \frac{n\sum xy - \left(\sum x\right)\left(\sum y\right)}{\sqrt{\left[n\sum x^2 - \left(\sum x\right)^2\right]\left[n\sum y^2 - \left(\sum y\right)^2\right]}}$$

$$r = \frac{10(658) - (55)(105)}{\sqrt{\left[10(385) - (55)^2\right]\left[10(1,221) - (105)^2\right]}}$$

$$r = \frac{805}{\sqrt{(825)(1,185)}} = 0.814$$

> **Stat Facts**
>
> The **coefficient of determination**, r^2, represents the percentage of the variation in y that is explained by the regression line.

The coefficient of determination becomes:

$$r^2 = (.0814)2 = 0.663$$

In other words, 66.3 percent of the variation in the number of items on the counter is explained by the Month variable. If $r^2 = 1$, all of the variation in y is explained by the variable x. If $r^2 = 0$, none of the variation in y is explained by the variable x.

Using Excel for Simple Regression

Now that we have burned out our calculators with all of these fancy equations, let me show you how Excel does it all for us.

1. Start by placing the bathroom counter data in Columns A and B in a blank sheet.

2. Go to Tools menu and select Data Analysis. (Refer to the section "Installing the Data Analysis Add-in" from Chapter 2 if you don't see the Data Analysis command on the Tools menu.)

3. From the Data Analysis dialog box, select Regression as shown in Figure 20.9 and click OK.

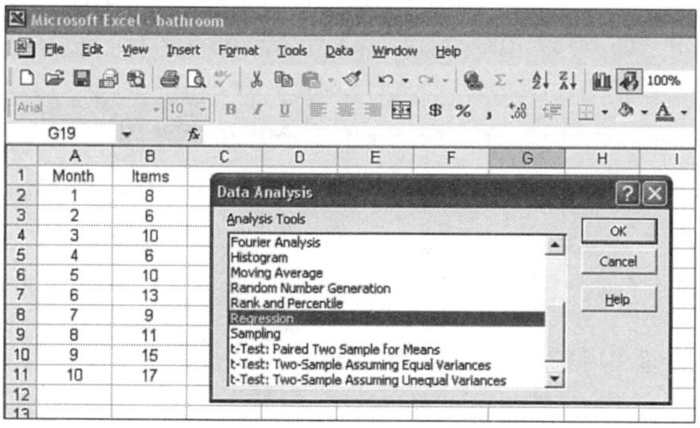

Figure 20.9

Setting up simple regression with Excel.

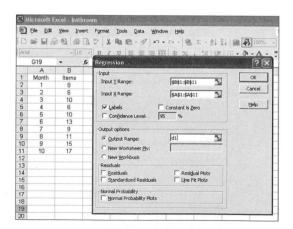

4. Set up the Regression dialog box according to Figure 20.10.

Figure 20.10

The Regression dialog box.

5. Click OK. Figure 20.11 shows the final regression results.

D	E	F	G	H	I	J
SUMMARY OUTPUT						
Regression Statistics						
Multiple R	0.814159943					
R Square	0.662856412					
Adjusted R Square	0.620713464					
Standard Error	2.234712374					
Observations	10					
ANOVA						
	df	*SS*	*MS*	*F*	*Significance F*	
Regression	1	78.54848485	78.54848485	15.72876214	0.004142356	
Residual	8	39.95151515	4.993939394			
Total	9	118.5				
	Coefficients	*Standard Error*	*t Stat*	*P-value*	*Lower 95%*	*Upper 95%*
Intercept	5.133333333	1.526599178	3.362594063	0.009893639	1.612987039	8.653679627
Month	0.975757576	0.246033735	3.965950345	0.004142356	0.408402399	1.543112753

Figure 20.11

Final results of the regression analysis in Excel.

These results are consistent with what we found after grinding it out in the previous sections. Because the *p*-value for the independent variable Month is shown as 0.00414, which is less than $\alpha = 0.01$, we can reject the null hypothesis and conclude that a relationship between the variables does exist. Deb has to believe me now!

A Simple Regression Example with Negative Correlation

Both examples in the chapter have involved a positive relationship between *x* and *y*. This example will summarize performing simple regression with a negative relationship.

Recently, I had the opportunity to "bond" with my son Brian as we shopped for his first car when he turned 16. Brian has visions of Mercedes and BMWs dancing in his head, whereas I was thinking more along the line of Hondas and Toyotas. After many "discussions" on the matter, we compromised on looking for 1999 Volkswagen Jettas. However, Brian had two requirements:

◆ It had to be black.

◆ It had to be the "new body" style.

Apparently, somebody at Volkswagen had the brilliant idea back in 1999 to subtly change the design of the Jetta halfway through the production year. Personally, I would never have noticed the difference. Brian, on the other hand, wouldn't be caught dead driving the original version, essentially eliminating half the used 1999 Volkswagen Jettas on the market. Undeterred, I searched the world over, asking each seller, "Is it the new body style?" Ahh, the joys of parenthood! Anyway, what follows is a table showing the mileage of 8 cars with the new body style and their asking price. The remainder of this chapter demonstrates the correlation and regression technique using this data.

Data for Car Example

Mileage	Price	Mileage	Price
21,800	$16,000	65,800	$10,500
34,000	$11,500	72,100	$12,300
41,700	$13,400	76,500	$8,200
53,500	$14,800	84,700	$9,500

The following table, which shows the data in thousands, will be used for the various equations

Mileage x	Price y	xy	x^2	y^2
21.8	16.0	348.80	475.24	256.00
34.0	11.5	391.00	1,156.00	132.25
41.7	13.4	558.78	1,738.89	179.56
53.5	14.8	791.80	2,862.25	219.04
65.8	10.5	690.90	4,329.64	110.25
72.1	12.3	886.83	5,198.41	151.29
76.5	8.2	627.30	5,852.25	67.24
84.7	9.5	804.65	7,174.09	90.25
$\sum x = 450.1$	$\sum y = 96.2$	$\sum xy = 5,100.1$	$\sum x^2 = 28,786.8$	$\sum y^2 = 1,205.9$

$$\bar{x} = \frac{450}{8} = 56.3 \qquad \bar{y} = \frac{96.2}{8} = 12.0$$

The correlation coefficient can be found using:

$$r = \frac{n\sum xy - \left(\sum x\right)\left(\sum y\right)}{\sqrt{\left[n\sum x^2 - \left(\sum x\right)^2\right]\left[n\sum y^2 - \left(\sum y\right)^2\right]}}$$

$$r = \frac{8(5,100.1) - (450.1)(96.2)}{\sqrt{\left[8(28,786.8) - (450)^2\right]\left[8(1,205.9) - (96.2)^2\right]}}$$

$$r = \frac{-2,498.82}{\sqrt{(27,794.4)(392.76)}} = -0.756$$

The negative correlation indicates that as mileage (x) increases, the price (y) decreases as we would expect. The coefficient of determination becomes:

$$r^2 = (-0.756)^2 = 0.572$$

Approximately 57 percent of the variation in price is explained by the variation in mileage. The regression line is determined using:

$$b = \frac{n\sum xy - \left(\sum x\right)\left(\sum y\right)}{n\sum x^2 - \left(\sum x\right)^2} = \frac{8(5100.1) - (450.1)(96.2)}{8(28786.8) - (450.1)^2}$$

$$b = \frac{-2498.82}{27704.39} = -0.0902$$

$$a = \bar{y} - b\bar{x} = 12.025 - (-0.0902)56.26 = 17.100$$

The regression line can be described by the equation:

$$\hat{y} = 17.1 - 0.0902x$$

This equation is shown graphically in Figure 20.12.

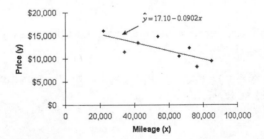

Figure 20.12

Regression line for car example.

What would the predicted price be for a car with 45,000 miles?

$$\hat{y} = 17.1 - 0.0902(45.0) = 13.041$$

The regression line would predict that a car with 45,000 miles would be priced at $13,041. What would be the 90 percent confidence interval at $x = 45000$? The standard error of the estimate would be:

$$s_e = \sqrt{\frac{\sum y^2 - a\sum y - b\sum xy}{n-2}}$$

$$s_e = \sqrt{\frac{(1205.9) - 17.1(96.2) - (-0.0902)(5100.1)}{8-2}}$$

$$s_e = \sqrt{\frac{(1205.9) - (1645.02) + (460.03)}{6}} = 1.867$$

The critical t-statistic for n − 2 = 8 − 2 = 6 degrees of freedom and a 90 percent confidence interval is t_c = 1.943 from Table 4 in Appendix B. Our confidence interval is then:

$$CI = \hat{y} \pm t_c s_e \sqrt{\frac{1}{n} + \frac{\left(x - \bar{x}\right)^2}{\left(\sum x^2\right) - \frac{\left(\sum x\right)^2}{n}}}$$

$$CI = 13.041 \pm (1.934)(1.867) \sqrt{\frac{1}{8} + \frac{(45 - 56.26)^2}{(28786.8) - \frac{(450.1)^2}{8}}}$$

$$CI = 13.04 \pm (1.934)(1.867)(0.402) = 13.04 \pm 1.452$$

$$CI = 11.589 \text{ and } 14.493$$

The 90 percent confidence interval for a car with 45,000 miles is between $11,589 and $14,493.

Is the relationship between mileage and price statistically significant at the α =0.10 level? Our hypotheses statement is:

H_0: β = 0

H_1: $\beta \neq 0$

The standard error of the slope, s_b, is found using:

$$s_b = \frac{s_e}{\sqrt{\sum x^2 - n\bar{x}^2}} = \frac{1.867}{\sqrt{28786.8 - 8(56.26)^2}}$$

$$s_b = \frac{1.867}{\sqrt{3465.3}} = 0.0317$$

The calculated test statistic for this hypothesis is:

$$t = \frac{b - \beta_{H_0}}{s_b} = \frac{-0.0902 - 0}{0.0317} = -2.845$$

The critical t-statistic is taken from the Students' t-distribution with $n - 2 = 8 - 2 = 6$ degrees of freedom. With a two-tail test and $\alpha = 0.10$, $t_c = 1.943$ according to Table 4 in Appendix B. Because $|t| > |t_c|$, we reject the null hypothesis and conclude there is a relationship between the mileage and price variable. We use the absolute values because the calculated t-statistic is in the left tail of the t-distribution with a two-tail hypothesis test.

Assumptions for Simple Regression

For all these results to be valid, we need to make sure that the underlying assumptions of simple regression are not violated. These assumptions are as follows:

- Individual differences between the data and the regression line, $\left(y_i - \hat{y}_i \right)$, are independent of one another.

- The observed values of y are normally distributed around the predicted value, $\hat{y}$.

- The variation of y around the regression line is equal for all values of x.

Unfortunately (or fortunately), the techniques to test these assumptions go beyond the level of this book.

Simple Versus Multiple Regression

Simple regression is limited to examining the relationship between a dependent variable and only one independent variable. If more than one independent variable is involved in the relationship, then we need to graduate to multiple regression. The regression equation for this method looks like:

$$\hat{y} = a + b_1 x_1 + b_2 x_2 + \ldots + b_n x_n$$

As you can imagine, this technique gets *really* messy and goes beyond the scope of this book. I'll have to save this topic for *The Complete Idiot's Guide to Statistics, Part 2*. Uh oh, I think I just heard Deb faint.

Your Turn

1. The following table shows the payroll for 10 Major League Baseball teams (in millions) for the 2002 season, along with the number of wins for that year.

Payroll	Wins	Payroll	Wins
$171	103	$56	62
$108	75	$62	84
$119	92	$43	78
$43	55	$57	73
$58	56	$75	67

Calculate the correlation coefficient. Test to see whether the correlation coefficient is not equal to 0 at the 0.05 level?

2. Using the data from Problem 1, answer the following questions:

   a) What is the regression line that best fits the data?

   b) Is the relationship between payroll and wins statistically significant at the 0.05 level?

   c) What is the predicted number of wins with a $70 million payroll?

   d) What is the 99 percent confidence interval around the mean number of wins for a $70 million payroll?

   e) What percent of the variation in wins is explained by the payroll?

The Least You Need to Know

♦ The independent variable (x) causes variation in the dependent variable (y).

♦ The correlation coefficient, r, indicates both the strength and direction of the relationship between the independent and dependent variables.

♦ The technique of simple regression enables us to describe a straight line that best fits a series of ordered pairs (x,y).

♦ The least squares method is a mathematical procedure to identify the linear equation that best fits a set of ordered pairs by finding values for a, the y-intercept, and b, the slope.

♦ The standard error of the estimate, s_e, measures the amount of dispersion of the observed data around the regression line.

♦ The coefficient of determination, r^2, represents the percentage of the variation in y that is explained by the regression line.

Solutions to "Your Turn"

Chapter 1

1. Inferential statistics, because it would not be feasible to survey every Asian American household in the country. These results would be based on a sample of the population and used to make an inference on the entire population.

2. Inferential statistics, because it would not be feasible to survey every household in the country. These results would be based on a sample of the population and used to make an inference on the entire population.

3. Descriptive statistics, because Barry Bonds's batting average is based on the entire population, which is every at-bat in his career.

4. Descriptive statistics, because the average SAT score would be based on the entire population, which is the 2002 freshman class.

5. Inferential statistics, because it would not be feasible to survey every American in the country. These results would be based on a sample of the population and used to make an inference on the entire population.

Chapter 2

1. Interval data, because temperature in degrees Fahrenheit does not contain a true zero point.

2. Ratio data, because monthly rainfall does have a true zero point.

3. Ordinal data, because a Master's degree is a higher level of education than a Bachelor's or high school degree. However, we cannot claim that a Master's degree is 2 or 3 times higher than the others.

4. Nominal data, because we cannot place the categories in any type of order.

5. Ratio data, because age does have a true zero point.

6. Definitely nominal data, unless you want to get into an argument about which is the lesser gender!

7. Interval data, because the difference between years is meaningful but a true zero point does not exist.

8. Nominal data, because I am not prepared to name one political party superior to another.

9. Nominal data, because these are simply unordered categories.

10. Ordinal data, because we can specify that "Above Expectation" is higher on the performance scale than the other 2 but we cannot comment on the differences between the categories.

11. Nominal data, because we cannot claim a person wearing the number "10" is any better then a person wearing the number "4".

12. Ordinal data, because we cannot comment on the difference in performance between students. The top 2 students may be very far apart grade-wise, whereas the second and third students could be very close.

13. Ratio data, because these exam scores have a true zero point.

14. Nominal data, because there is no order in the states' categories.

Chapter 3

1.

Exam Grade	Number of Students
56–60	2
61–65	1

Exam Grade	Number of Students
66–70	2
71–75	6
76–80	3
81–85	8
86–90	5
91–95	3
96–100	6

2.

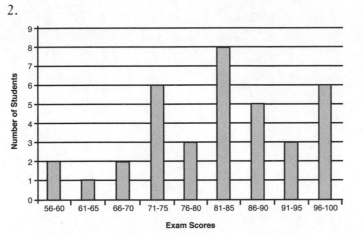

Histogram for Exam Grades.

3.

Exam Grade	Number of Students	Percentage	Cumulative Percentage
56–60	2	2/36 = .06	.06
61–65	1	1/36 = .03	.09
66–70	2	2/36 = .06	.15
71–75	6	6/36 = .17	.32
76–80	3	3/36 = .08	.40
81–85	8	8/36 = .22	.62
86–90	5	5/36 = .14	.76
91–95	3	3/36 = .08	.84
96–100	6	6/36 = .16	1.00
	Total = 36		

4.

Pie Chart for Exam Grades

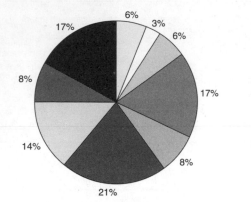

☐	56-60
☐	61-65
☐	66-70
■	71-75
■	76-80
■	81-85
☐	86-90
■	91-95
■	96-100

5.

Stem and Leaf for Problem 5

```
5|88999
6|0268
7|222455899
8|1123455566689
9|125688999
Stem and leaf for Problem 5
```

6.

Stem and Leaf for Problem 6

```
5 (5)|8
6 (0)|02
6 (5)|68
7 (0)|2224
7 (5)|55899
8 (0)|11234
8 (5)|55566689
9 (0)|12
9 (5)|5688999
Stem and leaf for Problem 6
```

Chapter 4

1. Mean = 15.9, Median = 17, Mode = 24

2. Mean = 81.7, Median = 82, Mode = 82

3. Mean = 32.7, Median = 32.5, Mode = 36

4. Mean = 7.2, Median = 6, Mode = 6

5. $\bar{x} = \dfrac{(22 \times 8) + (27 \times 37) + (32 \times 25) + (37 \times 48) + (42 \times 27) + (47 \times 10)}{8 + 37 + 25 + 48 + 27 + 10} = 34.5$ years old

6. $\bar{x} = \dfrac{(3 \times 118) + (2 \times 125) + (1 \times 107)}{3 + 2 + 1} = 118.5$

7. $\bar{x} = \dfrac{(5 \times 1) + (7 \times 2) + (10 \times 3) + (8 \times 4) + (12 \times 5) + (3 \times 6)}{5 + 7 + 10 + 8 + 12 + 3} = 3.5$ years of service

Chapter 5

1.

x_i	x^2
20	400
15	225
24	576
10	100
8	64
19	361
24	576
$\displaystyle\sum_{i=1}^{n} x_i = 120$	$\displaystyle\sum_{i=1}^{n} x_i^2 = 2302$

$$\left(\sum_{i=1}^{n} x_i\right)^2 = (120)^2 = 14,400$$

$$s^2 = \dfrac{\displaystyle\sum_{i=1}^{n} x_i^2 - \dfrac{\left(\displaystyle\sum_{i=1}^{n} x_i\right)^2}{n}}{n-1} = \dfrac{2,302 - \dfrac{14,400}{7}}{6} = 40.8$$

$$s = \sqrt{40.8} = 6.4$$

Range = 24 − 8 = 16

2.

x_i	x^2
84	7,056
82	6,724

continues

continued

x_i	x^2
90	8,100
77	5,929
75	5,625
77	5,929
82	6,724
86	7,396
82	6,724
$\sum\limits_{i=1}^{n} x = 735$	$\sum\limits_{i=1}^{n} x_i^2 = 60{,}207$

$$\left(\sum_{i=1}^{n} x_i\right)^2 = (735)^2 = 540{,}225$$

$$\sigma^2 = \frac{\sum\limits_{i=1}^{N} x_i^2 - \dfrac{\left(\sum\limits_{i=1}^{N} x_i\right)^2}{N}}{N} = \frac{60{,}207 - \dfrac{540{,}225}{9}}{9} = 20.2$$

$$\sigma = \sqrt{20.2} = 4.5$$

Range = 75 − 50 = 25

3. Range = 25, Variance = 75.4, Standard Deviation = 8.7

4. 2   5   6   6   *6*   *8*   10   11   11   15

   $Q_1 = 5.5$ $Q_2 = 7$ $Q_3 = 11$

Note the median of the data set is underlined.

5. $\bar{x} = \dfrac{\sum\limits_{i=1}^{m}(f_i \times x_i)}{\sum\limits_{i=1}^{m} f_i} = \dfrac{(8\times22)+(37\times27)+(25\times32)+(48\times37)+(27\times42)+(10\times47)}{8+37+25+48+27+10} = 34.5$

x_i	f_i	$\bar{x}$	$\left(x_i - \bar{x}\right)$	$\left(x_i - \bar{x}\right)^2$	$\left(x_i - \bar{x}\right)^2 f_i$
22	8	34.5	−12.5	156.25	1,250.00
27	37	34.5	−7.5	56.25	2,081.25
32	25	34.5	−2.5	6.25	156.25

x_i	f_i	$\bar{x}$	$(x_i - \bar{x})$	$(x_i - \bar{x})^2$	$(x_i - \bar{x})^2 f_i$
37	48	34.5	2.5	6.25	300.00
42	27	34.5	7.5	56.25	1,518.75
47	10	34.5	12.5	156.25	1,562.50

$$n = \sum_{i=1}^{m} f_i = 155 \qquad\qquad \sum_{i=1}^{m} (x_i - \bar{x})^2 f_i = 6,868.75$$

$$s = \sqrt{\frac{\sum_{i=1}^{m}(x_i - \bar{x})^2 f_i}{n-1}} = \sqrt{\frac{6,868.75}{155-1}} = \sqrt{44.60} = 6.68 \ \text{years}$$

6. $\bar{x} = \dfrac{\sum_{i=1}^{m}(f_i \times x_i)}{\sum_{i=1}^{m} f_i} = \dfrac{(5\times1)+(7\times2)+(10\times3)+(8\times4)+(12\times5)+(3\times6)}{5+7+10+8+12+3} = 3.5$

x_i	f_i	$\bar{x}$	$(x_i - \bar{x})$	$(x_i - \bar{x})^2$	$(x_i - \bar{x})^2 f_i$
1	5	3.5	−2.5	6.25	31.25
2	7	3.5	−1.5	2.25	15.75
3	10	3.5	−0.5	0.25	2.50
4	8	3.5	0.5	0.25	2.00
5	12	3.5	1.5	2.25	27.00
6	3	3.5	2.5	6.25	18.75

$$n = \sum_{i=1}^{m} f_i = 45 \qquad\qquad (x_i - \bar{x})^2 f_i = 97.25$$

$$s = \sqrt{\frac{\sum_{i=1}^{m}(x_i - \bar{x})^2 f_i}{n-1}} = \sqrt{\frac{97.25}{45-1}} = \sqrt{2.21} = 1.49 \ \text{years}$$

7. Using the empirical rule, 95 percent of the values will fall within $k = 2$ standard deviations from the mean.

$$\mu = k\sigma = 75 + 2(10) = 95, \qquad \mu = k\sigma = 75 - 2(10) = 55$$

Therefore, 95 percent of the data values should fall between 55 and 95.

8. The values 38 and 62 are 2 standard deviations from the mean of 50. This can be shown with the following:

$$\mu + k\sigma = 62, \qquad\qquad \mu + k\sigma = 38$$

$$k = \frac{62 - \mu}{\sigma} = \frac{62 - 50}{6} = 2.0, \qquad k = -\left(\frac{38 - \mu}{\sigma}\right) = -\left(\frac{38 - 50}{6}\right) = 2.0$$

Using Chebyshev's theorem, at least $\left(1 - \dfrac{1}{k^2}\right) \times 100\% = \left(1 - \dfrac{1}{2^2}\right) \times 100\% = 75$ per-cent of the data values will fall between 38 and 62.

Chapter 6

1a. Empirical, because we have historical data for Sammy Sosa's batting average.

1b. Classical, because we know the number of cards and the number of Aces in the deck.

1c. If I have data from my last several rounds of golf, this would be empirical, otherwise subjective.

1d. Classical, because we can calculate the probability based on the lottery rules.

1e. Subjective, because I would not be collecting data for this experiment.

1f. Subjective, because I would not be collecting data for this experiment.

2a. Yes.

2b. No, probability cannot be greater than 1.

2c. No, probability cannot be greater than 100 percent.

2d. No, probability cannot be less than 1.

2e. Yes.

2f. Yes.

3a. $P[A] = \dfrac{52}{125} = 0.42$

3b. $P[B] = \dfrac{41}{125} = 0.33$

3c. $P[A \text{ and } B] = P[A \cap B] = \dfrac{23}{125} = 0.18$

3d. The following table identifies the total number of families for the union of Events A and B.

Race	Internet	Number of Families
Asian American	Yes	23
Asian American	No	18
Caucasian	Yes	15
African American	Yes	14
		Total = 70

$$P[A \text{ or } B] = P[A \cup B] = \frac{70}{125} = 0.56$$

Chapter 7

1. $P[A] = \dfrac{177}{260} = 0.68$

2. $P[B] = \dfrac{152}{260} = 0.58$

3. $P[A'] = \dfrac{83}{260} = 0.32$

4. $P[B'] = \dfrac{108}{260} = 0.42$

5. $P[A / B] = \dfrac{98}{152} = 0.64$

6. $P[A' / B] = \dfrac{54}{152} = 0.36$

7. $P[A / B'] = \dfrac{79}{108} = 0.73$

8. $P[A \text{ and } B] = P[A / B] \cdot P[B] = (0.64) \cdot (0.58) = 0.37$

9. $P[A \text{ and } B'] = P[A / B'] \cdot P[B'] = (0.73) \cdot (0.42) = 0.31$

10. $P[A \text{ or } B] = P[A] + P[B] - P[A \text{ and } B] = 0.68 + 0.58 - 0.37 = 0.89$

11. $P[A \text{ or } B'] = P[A] + P[B'] - P[A \text{ and } B'] = 0.68 + 0.42 - 0.31 = 0.79$

12. $P[B / A] = \dfrac{P[B] \cdot P[A / B]}{(P[B] \cdot P[A / B]) + (P[B'] P[A / B'])}$

$$P[B / A] = \frac{(0.58) \cdot (0.64)}{(0.58) \cdot (0.64) + (0.42) \cdot (0.73)} = \frac{0.37}{0.37 + 0.31} = 0.54$$

Chapter 8

1. $3 \times 8 \times 4 \times 3 = 288$ different meals

2. There are $4 \times 4 \times 4 \times 4 \times 4 \times 4 \times 4 \times 4 \times 4 \times 4 = 1{,}048{,}576$ different ways to answer the exam. If only one of these sequences is correct, the probability is $1/1048576 = 0.00000095$ that the student will correctly guess the correct sequence.

3. $13! = 6{,}227{,}020{,}800$ different ordered arrangements

4. $_8P_3 = \dfrac{8!}{(8-3)!} = 8 \cdot 7 \cdot 6 = 336$

5. $_{10}P_2 = \dfrac{10!}{(10-2)!} = 10 \cdot 9 = 90$

6. $_{40}P_3 = \dfrac{40!}{(40-3)!} = 40 \cdot 39 \cdot 38 = 59{,}280$

7. $_{12}C_3 = \dfrac{12!}{(12-3)!3!} = \dfrac{12 \cdot 11 \cdot 10}{3 \cdot 2 \cdot 1} = 220$

8. $_{50}C_{12} = \dfrac{50!}{(50-12)!12!} = \dfrac{50 \cdot 49 \cdot 48 \cdot 47 \cdot 46 \cdot 45 \cdot 44 \cdot 43 \cdot 42 \cdot 41 \cdot 40 \cdot 39}{12 \cdot 11 \cdot 10 \cdot 9 \cdot 8 \cdot 7 \cdot 6 \cdot 5 \cdot 4 \cdot 3 \cdot 2 \cdot 1} = 121{,}399{,}651{,}100$

9.

Number of Cats	Number of Families x_i	Probability $P[x_i]$	x_i^2	$x_i^2 \cdot P[x_i]$
0	137	137/450 = 0.304	0	0
1	160	160/450 = 0.356	1	0.356
2	112	112/450 = 0.249	4	0.996
3	31	31/450 = 0.069	9	0.621
4	10	10/450 = 0.022	16	0.352
Total				$\displaystyle\sum_{i=1}^{n} x_i^2 \cdot P[x_i] = 2.325$

$$\mu = \sum_{i=1}^{n} x_i \cdot P[x_i] = (0 \times 0.304) + (1 \times 0.356) + (2 \times 0.249) + (3 \times 0.069) + (4 \times 0.022) = 1.149$$

$$\sigma^2 = \left(\sum_{i=1}^{n} x_i \cdot P[x_i]\right) - \mu^2 = 2.325 - (1.149)^2 = 1.005$$

$$\sigma = \sqrt{\sigma^2} = \sqrt{1.005} = 1.002$$

Chapter 9

1. Because $n = 10$, $r = 7$, $p = 0.5$

$$P[7,10] = \frac{10!}{(10-7)!7!} \cdot 0.5^7 \cdot 0.5^{10-7} = \frac{10 \cdot 9 \cdot 8 \cdot 7 \cdot 6 \cdot 5 \cdot 4}{7 \cdot 6 \cdot 5 \cdot 4 \cdot 3 \cdot 2 \cdot 1}(0.0078)(0.125) = 0.117$$

The binomial table in Appendix B can also be used here.

2. Because $n = 6$, $r = 3$, $p = 0.75$

$$P[3,6] = \frac{6!}{(6-3)!3!} \cdot 0.75^3 \cdot 0.25^{6-3} = \frac{6 \cdot 5 \cdot 4}{3 \cdot 2 \cdot 1}(0.4219)(0.0156) = 0.1316$$

3. The probability of making at least 6 of 8 is $P[6,8] + P[7,8] + P[8,8]$. Because $n = 8$, $p = 0.8$

$$P[6,8] = \frac{8!}{(8-6)!6!} \cdot 0.8^6 \cdot 0.2^{8-6} = \frac{8 \cdot 7}{2 \cdot 1}(0.2621)(0.04) = 0.2936$$

$$P[7,8] = \frac{8!}{(8-7)!7!} \cdot 0.8^7 \cdot 0.2^{8-7} = (8)(0.2097)(0.2) = 0.3355$$

$$P[8,8] = \frac{8!}{(8-8)!8!} \cdot 0.8^8 \cdot 0.2^{8-8} = (1)(0.1678)(1) = 0.1678$$

Therefore, the probability of making at least 6 out of 8 is $0.2936 + 0.3355 + 0.1678 = 0.7969$.

4. Because $n = 12$, $r = 6$, $p = 0.2$

$$P[6,12] = \frac{12!}{(12-6)!6!} \cdot 0.2^6 \cdot 0.8^{12-6} = \frac{12 \cdot 11 \cdot 10 \cdot 9 \cdot 8 \cdot 7}{6 \cdot 5 \cdot 4 \cdot 3 \cdot 2 \cdot 1}(0.000064)(0.2621) = 0.0155$$

5. The probability of no more than 2 out of the next 7 is $P[0,7] + P[1,7] + P[2,7]$. Because $n = 7$, $p = 0.05$

$$P[0,7] = \frac{7!}{(7-0)!0!} \cdot 0.05^0 \cdot 0.95^{7-0} = (1)(1)(0.04) = 0.6983$$

$$P[1,7] = \frac{7!}{(7-1)!1!} \cdot 0.05^1 \cdot 0.95^{7-1} = (7)(0.05)(0.7351) = 0.2573$$

$$P[2,7] = \frac{7!}{(7-2)!2!} \cdot 0.05^2 \cdot 0.95^{7-2} = \frac{7 \cdot 6}{2 \cdot 1} \cdot (0.0025)(0.7738) = 0.0406$$

Therefore, the probability that no more than 2 of the next 7 people will purchase is $0.6983 + 0.2573 + 0.0406 = 0.9962$.

6. Because $n = 5$, $p = 0.37$

$$P[0,5] = \frac{5!}{(5-0)!0!} \cdot 0.37^0 \cdot 0.63^{5-0} = (1)(1)(0.0992) = 0.0992$$

$$P[1,5] = \frac{5!}{(5-1)!1!} \cdot 0.37^1 \cdot 0.63^{5-1} = (5)(0.37)(0.1575) = 0.2914$$

$$P[2,5] = \frac{5!}{(5-2)!2!} \cdot 0.37^2 \cdot 0.63^{5-2} = \left(\frac{5 \cdot 4}{2 \cdot 1}\right)(0.1369)(0.2500) = 0.3423$$

$$P[3,5] = \frac{5!}{(5-3)!3!} \cdot 0.37^3 \cdot 0.63^{5-3} = \left(\frac{5 \cdot 4 \cdot 3}{3 \cdot 2 \cdot 1}\right)(0.0507)(0.3969) = 0.2010$$

$$P[4,5] = \frac{5!}{(5-4)!4!} \cdot 0.37^4 \cdot 0.63^{5-4} = (5)(0.0187)(0.63) = 0.0590$$

$$P[5,5] = \frac{5!}{(5-5)!5!} \cdot 0.37^5 \cdot 0.63^{5-5} = (1)(0.0069)(1) = 0.0069$$

r	P[r,5]
0	0.0992
1	0.2914
2	0.3423
3	0.2010
4	0.0590
5	0.0069
	Total = 1.0

Chapter 10

1. $P[4] = \dfrac{\left(6^4\right)\left(2.71838^{-6}\right)}{4!} = \dfrac{(1,296)(.002479)}{24} = 0.1339$

2. $P[5] = \dfrac{\left(7.5^5\right)\left(2.71838^{-7.5}\right)}{5!} = \dfrac{(23,730.469)(0.0005531)}{120} = 0.1094$

3. $P[x > 2] = 1 - P[x \leq 2] = 1 - (P[x = 0] + P[x = 1] + P[x = 2])$

$$P[0] = \frac{\left(4.2^0\right)\left(2.71838^{-4.2}\right)}{0!} = \frac{(1)(.0150)}{1} = 0.150$$

$$P[1] = \frac{\left(4.2^1\right)\left(2.71838^{-4.2}\right)}{1!} = \frac{(4.2)(.0150)}{1} = 0.630$$

$$P[2] = \frac{(4.2^2)(2.71838^{-4.2})}{2!} = \frac{(17.64)(.0150)}{2} = 0.1323$$

$$P[x > 2] = 1 - (0.0150 + 0.0630 + 0.1323) = 0.7897$$

4. $P[x \le 3] = (P[x = 0] + P[x = 1] + P[x = 2] + P[x = 3])$

$$P[0] = \frac{(3.6^0)(2.71838^{-3.6})}{0!} = \frac{(1)(.027324)}{1} = 0.0273$$

$$P[1] = \frac{(3.6^1)(2.71838^{-3.6})}{1!} = \frac{(3.6)(.027324)}{1} = 0.0984$$

$$P[2] = \frac{(3.6^2)(2.71838^{-3.6})}{2!} = \frac{(12.96)(.027324)}{2} = 0.1771$$

$$P[3] = \frac{(3.6^3)(2.71838^{-3.6})}{3!} = \frac{(46.656)(.027324)}{6} = 0.2125$$

$$P[x \le 3] = (0.0273 + 0.0984 + 0.1771 + 0.2125) = 0.5152$$

5. $P[1] = \dfrac{(2.5^1)(2.71838^{-2.5})}{1!} = \dfrac{(2.5)(.082085)}{1} = 0.2052$

6. $P[x] = \dfrac{(np)^x \cdot e^{-(np)}}{x!}$, $n = 25$, $p = 0.05$, $np = 1.25$

$$P[2] = \frac{(1.25)^2 \cdot e^{-(1.25)}}{2!} = \frac{(1.5625)(0.286505)}{2} = 0.2238$$

Chapter 11

1a. $z_{65.5} = \dfrac{65.5 - 62.6}{3.7} = +0.78$, $P[z > +0.78] = 1 - P[z \le +0.78] = 1 - 0.7834 = 0.2166$

1b. $z_{58.1} = \dfrac{58.1 - 62.6}{3.7} = -1.22$, $P[z > -1.22] = P[z \le +1.22] = 0.8880$

1c. $z_{70} = \dfrac{70 - 62.6}{3.7} = +2.0$, $z_{61} = \dfrac{61 - 62.6}{3.7} = -0.43$,

$P[-0.43 \le z \le +2.0] = P[z \le +2.0] - P[z \le -0.43]$,

$P[-0.43 \le z \le +2.0] = 0.9772 - 0.3327 = 0.6445$

2a. $z_{190} = \dfrac{190 - 176}{22.3} = +0.63, \quad P[z \le +0.63] = 0.7349$

2b. $z_{158} = \dfrac{158 - 176}{22.3} = -0.81, \quad P[z \le -0.81] = 1 - P[z \le +0.81] = 0.2098$

2c. $z_{168} = \dfrac{168 - 176}{22.3} = -0.36, \quad z_{150} = \dfrac{150 - 176}{22.3} = -1.17$

$P[-1.17 \le z \le -0.36] = P[z \le -0.36] - P[z \le -1.17]$

$P[z \le -0.36] = 1 - P[z \le +0.36] = 0.3599, \quad P[z \le -1.17] = 1 - P[z \le +1.17] = 0.1218$

$P[-1.17 \le z \le -0.36] = 0.3599 - 0.1218 = 0.2381$

3a. $z_{31} = \dfrac{31 - 37.5}{7.6} = -0.86, \quad P[z > -0.86] = P[z \le +0.86] = 0.8038$

3b. $z_{42} = \dfrac{42 - 37.5}{7.6} = +0.59, \quad P[z \le +0.59] = 0.7231$

3c. $z_{45} = \dfrac{45 - 37.5}{7.6} = +0.99, \quad z_{40} = \dfrac{40 - 37.5}{7.6} = +0.33$

$P[+0.99 \le z \le +0.33] = P[z \le +0.99] - P[z \le +0.33] = 0.8381 - 0.6289 = 0.2092$

4. For this problem, $n = 14$, $p = 0.5$, and $q = 0.5$. We can use the normal approximation because $np = nq = (14)(0.5) = 7$. The binomial probabilities from the binomial table are $P[r = 4, 5, \text{ or } 6] = 0.0611 + 0.1222 + 0.1833 = 0.3666$. Also, $\mu = np = (14)(0.5) = 7$ and $\sigma = \sqrt{npq} = \sqrt{(14)(0.5)(0.5)} = 1.871$. The normal approximation would be finding $P[3.5 \le x \le 6.5]$.

$z_{6.5} = \dfrac{6.5 - 7}{1.871} = -0.27, \quad z_{3.5} = \dfrac{3.5 - 7}{1.871} = -1.87$

$P[-1.87 \le z \le -0.27] = P[z \le -0.27] - P[z \le -1.87]$

$P[z \le -0.27] = 1 - P[z \le +0.27] = 0.3946, \quad P[z \le -1.87] = 1 - P[z \le +1.87] = 0.0307$

$P[-1.87 \le z \le -0.27] = 0.3946 - 0.0307 = 0.3639$

Chapter 12

1. $k = \dfrac{N}{n} = \dfrac{75000}{500} = 150$

2. If every employee belonged to a particular department, certain departments could be chosen for the survey, with every individual in those departments asked to participate. Other answers are also possible.

3. If each employee can be classified as either a manager or a non-manager, ensure that the sample proportion for each type is similar to the proportion of managers and non-managers in the company. Other answers are also possible.

Chapter 13

1a. $\sigma_{\bar{x}} = \dfrac{\sigma}{\sqrt{n}} = \dfrac{10}{\sqrt{15}} = 2.58$

1b. $\sigma_{\bar{x}} = \dfrac{\sigma}{\sqrt{n}} = \dfrac{4.7}{\sqrt{12}} = 1.36$

1c. $\sigma_{\bar{x}} = \dfrac{\sigma}{\sqrt{n}} = \dfrac{7}{\sqrt{20}} = 1.57$

2a. $\sigma_{\bar{x}} = \dfrac{\sigma}{\sqrt{n}} = \dfrac{7.5}{\sqrt{9}} = 2.5$, $z_{17} = \dfrac{17-16}{2.5} = +0.40$, $P[z \le +0.40] = 0.6554$

2b. $z_{18} = \dfrac{18-16}{2.5} = +0.80$, $P[z > +0.80] = 1 - P[z \le +0.80] = 1 - 0.7881 = 0.2119$

2c. $z_{14.5} = \dfrac{14.5-16}{2.5} = -0.60$, $z_{16.5} = \dfrac{16.5-16}{2.5} = +0.20$

$P[z \le +0.20] = 0.5793$, $P[z \le -0.60] = 1 - P[z \le +0.60] = 0.2743$

$P\left[14.5 \le \bar{x} \le 16.5\right] = P[-0.60 \le z \le 0.20] = 0.5793 - 0.2743 = 0.3050$

3a. $\sigma_p = \sqrt{\dfrac{p(1-p)}{n}} = \sqrt{\dfrac{0.25(1-0.25)}{200}} = 0.0306$

3b. $\sigma_p = \sqrt{\dfrac{p(1-p)}{n}} = \sqrt{\dfrac{0.42(1-0.42)}{100}} = 0.0494$

3c. $\sigma_p = \sqrt{\dfrac{p(1-p)}{n}} = \sqrt{\dfrac{0.06(1-0.06)}{175}} = 0.0179$

4a. $\sigma_p = \sqrt{\dfrac{p(1-p)}{n}} = \sqrt{\dfrac{0.32(1-0.32)}{160}} = 0.0369$, $z_{0.30} = \dfrac{0.30-0.32}{0.0369} = -0.54$

$P[z \le -0.54] = 1 - P[z \le +0.54] = 0.2946$

4b. $z_{0.36} = \dfrac{0.36-0.32}{0.0369} = +1.08$, $P[z > +1.08] = 1 - P[z \le +1.08] = 1 - 0.8599 = 0.1401$

4c. $z_{0.29} = \dfrac{0.29 - 0.32}{0.0369} = -0.81$, $z_{0.37} = \dfrac{0.37 - 0.32}{0.0369} = +1.36$

$P[z \le +1.36] = 0.9131$, $P[z \le -0.81] = 1 - P[z \le +0.81] = 0.2090$

$P[.029 \le p_s \le 0.37] = P[-0.81 \le z \le +1.36] = 0.9131 - 0.2090 = 0.7041$

Chapter 14

1. $\sigma_{\bar{x}} = \dfrac{\sigma}{\sqrt{n}} = \dfrac{7.6}{\sqrt{40}} = 1.20$, $z_c = 2.17$

   Upper limit $= \bar{x} + z_c \sigma_{\bar{x}} = 31.3 + 2.17(1.20) = 33.90$

   Lower limit $= \bar{x} - z_c \sigma_{\bar{x}} = 31.3 - 2.17(1.20) = 28.70$

2. $n = \left(\dfrac{z\sigma}{E}\right)^2 = \left(\dfrac{(2.33)(15)}{5}\right)^2 = 48.9 \approx 49$

3. This is a trick question! The sample size is too small to be used from a population that is not normally distributed. This question goes beyond the scope of this book. You would need to consult a statistician.

4. Using Excel, we can calculate $\bar{x} = 13.9$ and $s = 6.04$.

   $\sigma^\mu_{\bar{x}} = \dfrac{s}{\sqrt{n}} = \dfrac{6.04}{\sqrt{30}} = 1.10$

   Upper limit $= \bar{x} + z_c \sigma^\mu_{\bar{x}} = 13.9 + 1.64(1.10) = 15.70$

   Lower limit $= \bar{x} - z_c \sigma^\mu_{\bar{x}} = 13.9 - 1.64(1.10) = 12.10$

5. Using Excel, we can calculate $\bar{x} = 46.92$.

   $\sigma = 12.7$, $z_c = 1.88$, and $\sigma_{\bar{x}} = \dfrac{\sigma}{\sqrt{n}} = \dfrac{12.7}{\sqrt{12}} = 3.67$

   Upper limit $= \bar{x} + z_c \sigma_{\bar{x}} = 46.92 + 1.88(3.67) = 53.82$

   Lower limit $= -z_c \sigma_{\bar{x}} = 46.92 - 1.88(3.67) = 40.02$

6. Using Excel, we can calculate $\bar{x} = 119.64$ and $s = 11.29$

   $\sigma^\mu_{\bar{x}} = \dfrac{s}{\sqrt{n}} = \dfrac{11.29}{\sqrt{11}} = 3.40$

   For a 98 percent confidence interval with $n - 1 = 11 - 1 = 10$ degrees of freedom, $t_c = 2.764$

   Upper limit $= \bar{x} + t_c \sigma^\mu_{\bar{x}} = 119.64 + 2.764(3.40) = 129.04$

   Lower limit $= \bar{x} - t_c \sigma^\mu_{\bar{x}} = 119.64 - 2.764(3.40) = 110.24$

7. This is another trick question! The sample size is too small to be used from a population that is not normally distributed. This question goes beyond the scope of this book. You would need to consult a statistician.

8. $p_s = \dfrac{11}{200} = 0.055$. Because $np_s = (200)(0.055) = 11$ and $nq_s = (200)(0.945) = 189$, we can use the normal approximation.

$$\sigma^{\mu}_p = \sqrt{\dfrac{p_s(1-p_s)}{n}} = \sqrt{\dfrac{(0.055)(0.945)}{200}} = 0.0161 \, , z_c = 1.96$$

Upper limit $= p_s + z_c \sigma_p = 0.055 + 1.96(0.0161) = 0.087$

Lower limit $= p_s - z_c \sigma_p = 0.055 - 1.96(0.0161) = 0.023$

9. $n = pq\left(\dfrac{z_c}{E}\right)^2 = (0.55)(0.45)\left(\dfrac{2.05}{0.04}\right)^2 = 650$

Chapter 15

1. $H_0: \mu = 1.7$

   $H_1: \mu \neq 1.7$

   $n = 35$, $\sigma = 0.5$ cups, $\sigma_{\bar{x}} = \dfrac{\sigma}{\sqrt{n}} = \dfrac{0.50}{\sqrt{35}} = 0.0845$ cups, $z_c \pm 1.64$

   Upper limit $= \mu_{H_0} + z_c \sigma_{\bar{x}} = 1.7 + (1.64)(0.0845) = 1.84$ cups

   Lower limit $= \mu_{H_0} - z_c \sigma_{\bar{x}} = 1.7 - (1.64)(0.0845) = 1.56$ cups

   Because $\bar{x} = 1.95$ cups, we reject H_0 and conclude that the population mean is not 1.7 cups per day.

2. $H_0: \mu \geq 40$

   $H_1: \mu < 40$

   $n = 50$, $\sigma = 12.5$ years, $\sigma_{\bar{x}} = \dfrac{\sigma}{\sqrt{n}} = \dfrac{12.5}{\sqrt{50}} = 1.768$ years,

   $z_c = -1.64$ (left tail of the distribution)

   Lower limit $= \mu_{H_0} + z_c \sigma_{\bar{x}} = 40 + (-1.64)(1.768) = 37.1$ years

   Because $\bar{x} = 38.7$ years, we do not reject H_0 and conclude that we do not have enough evidence to support the claim that the average age is less than 40 years old.

3. $H_0: \mu \leq 1000$

   $H_1: \mu > 1000$

$n = 32$, $\sigma = 325$ hours, $\sigma_{\bar{x}} = \dfrac{\sigma}{\sqrt{n}} = \dfrac{325}{\sqrt{32}} = 57.45$ hours,

$z_c = +2.05$ (right tail of the distribution)

Upper limit $= \mu_{H_0} + z_c\sigma_{\bar{x}} = 1000 + (2.05)(57.45) = 1117.8$ hours

Because $\bar{x} = 1190$ hours, we reject H_0 and conclude the average light bulb life exceeds 1,000 hours.

4. H_0: $\mu \geq 30$

   H_1: $\mu < 30$

   $n = 42$, $\sigma = 8.0$ minutes, $\sigma_{\bar{x}} = \dfrac{\sigma}{\sqrt{n}} = \dfrac{8.0}{\sqrt{42}} = 1.23$ minutes,

   $z_c = -2.33$ (left tail of the distribution)

   Lower limit $= \mu_{H_0} + z_c\sigma_{\bar{x}} = 30 + (-2.33)(1.23) = 27.13$ minutes

   Because $\bar{x} = 26.9$ minutes, we reject H_0 and conclude that the average delivery time is less than 30 minutes.

Chapter 16

1. H_0: $\mu = 1100$

   H_1: $\mu \neq 1100$

   $n = 70$, $\sigma = 310$, $\sigma_{\bar{x}} = \dfrac{\sigma}{\sqrt{n}} = \dfrac{310}{\sqrt{70}} = 37.05$, $z_c = \pm 1.64$,

   $z = \dfrac{\bar{x} - \mu_{H_0}}{\sigma_{\bar{x}}} = \dfrac{1035 - 1100}{37.05} = -1.75$,

   p-value $= (2)(P[z \leq -1.75]) = (2)(1 - P[z \leq +1.75]) = (2)(1 - 0.9599) = 0.0802$

   Because p-value $= \alpha$, we reject H_0 and conclude that the average SAT score does not equal 1,100.

2. H_0: $\mu \leq 35$

   H_1: $\mu > 35$

   $\bar{x} = 37.9$, $s = 6.74$, $n = 10$, $d.f. = n - 1 = 9$, $\hat{\sigma}_{\bar{x}} = \dfrac{s}{\sqrt{n}} = \dfrac{6.74}{\sqrt{10}} = 2.13$, $t_c = \pm 2.821$,

   $t = \dfrac{\bar{x} - \mu_{H_0}}{\hat{\sigma}_{\bar{x}}} = \dfrac{37.9 - 35}{2.13} = +1.36$

   Because $t \leq t_c$, we do not reject H_0 and conclude that the average class size equals 35 students.

3. $H_0: \mu \leq 7$

   $H_1: \mu > 7$

   $\bar{x} = 8.2$, $s = 4.29$, $n = 10$, $d.f. = n - 1 = 9$, $\hat{\sigma}_{\bar{x}} = \dfrac{s}{\sqrt{n}} = \dfrac{4.29}{\sqrt{30}} = 0.78$, $z_c = +1.64$

   $z = \dfrac{\bar{x} - \mu_{H_0}}{\hat{\sigma}_{\bar{x}}} = \dfrac{8.2 - 7}{0.78} = +1.54,$

   p-value $= P[z > +1.54] = 1 - P[z \leq +1.54] = 1 - 0.9382 = 0.0618$

   Because $z \leq z_c$ or p-value $> \alpha$, we do not reject H_0 and conclude that average gasoline consumption in the United States does not exceed 7 liters per car per day.

4. $H_0: p \geq 0.40$

   $H_1: p < 0.40$

   $\sigma_p = \sqrt{\dfrac{p_{H_0}(1 - p_{H_0})}{n}} = \sqrt{\dfrac{(0.40)(1 - 0.40)}{175}} = 0.037$, $z = \dfrac{\bar{p} - p_{H_0}}{\sigma_p} = \dfrac{0.30 - 0.40}{0.037} = -2.70$

   p-value $= P[z \leq -2.70] = 1 - P[z \leq +2.70] = 1 - 0.9965 = 0.0035$, $z_c = -2.33$

   Because p-value $\leq \alpha$, we reject H_0 and conclude that the proportion of Republicans is less than 40 percent.

Chapter 17

1. $H_0: \mu_1 = \mu_2$

   $H_1: \mu_1 \neq \mu_2$, Pennsylvania = 1, Ohio = 2

   $\sigma_{\bar{x}_1 - \bar{x}_2} = \sqrt{\dfrac{\sigma_1^2}{n_1} + \dfrac{\sigma_2^2}{n_2}} = \sqrt{\dfrac{(105)^2}{45} + \dfrac{(114)^2}{38}} = 24.22$, $z_c = \pm 1.96$

   $z = \dfrac{(\bar{x}_1 - \bar{x}_2) - (\mu_1 - \mu_2)_{H_0}}{\sigma_{\bar{x}_1 - \bar{x}_2}} = \dfrac{(552 - 530) - 0}{24.22} = +0.91$

   Because $z \leq z_c$, we do not reject H_0 and conclude there is not enough evidence to support a difference between the 2 states.

   p-value $= (2)P[z > +0.91] = (2)(1 - P[z \leq +0.91]) = 2(1 - 0.8186) = 0.3628$

2. $H_0: \mu_1 = \mu_2$

$H_1: \mu_1 \neq \mu_2$, $\bar{x}_1 = 88.3$, $s_1 = 7.30$, $\bar{x}_2 = 82.4$, $s_2 = 6.74$,

$$s_p = \sqrt{\frac{(n_1 - 1)s_1^2 + (n_2 - 1)s_2^2}{n_1 + n_2 - 2}} = \sqrt{\frac{(10)(7.30)^2 + (9)(6.74)^2}{11 + 10 - 2}} = 7.04$$

$$\sigma^\mu_{\bar{x}_1 - \bar{x}_2} = s_p \sqrt{\frac{1}{n_1} + \frac{1}{n_2}} = (7.04)\sqrt{\frac{1}{11} + \frac{1}{10}} = (7.04)\sqrt{0.1909} = 3.08$$

$$t = \frac{(\bar{x}_1 - \bar{x}_2) - (\mu_1 - \mu_2)_{H_0}}{\sigma^\mu_{\bar{x}_1 - \bar{x}_2}} = \frac{(88.3 - 82.4) - 0}{3.08} = +1.92$$

$d.f. = n_1 + n_2 - 2 = 11 + 10 - 2 = 19$, $t_c = \pm 1.729$

Because $t > t_c$, we reject H_0 and conclude the satisfaction scores are not equal between the 2 stores.

3. $H_0: \mu_d \leq 15$

$H_0: \mu_d \leq 15$, $\sum d = 21 + 23 + 11 + 19 + 15 + 20 + 17 + 23 + 17 = 166$

$\sum d^2 = 441 + 529 + 121 + 361 + 225 + 400 + 289 + 529 + 289 = 3,184$

$$\bar{d} = \frac{\sum d}{n} = \frac{166}{9} = 18.44,$$

$$s_d = \sqrt{\frac{\sum d^2 - \frac{(\sum d)^2}{n}}{n - 1}} = \sqrt{\frac{(3184) - \frac{(166)^2}{9}}{8}} = \sqrt{\frac{122.22}{8}} = 3.91$$

$$t = \frac{\bar{d} - \mu_d}{\frac{s_d}{\sqrt{n}}} = \frac{18.44 - 15}{\frac{3.91}{\sqrt{9}}} = \frac{3.44}{1.30} = +2.64, \ d.f. = n - 1 = 9 - 1 = 8, \ t_c = +1.860$$

Because $t > t_c$, we reject H_0 and conclude the weight-loss program claim is valid.

4. $H_0: p_1 \leq p_2$

$H_1: p_1 > p_2$, Population 1 = Florida, Population 2 = Nation

$$\bar{p}_1 = \frac{x_1}{n_1} = \frac{272}{400} = 0.68, \ \bar{p}_2 = \frac{x_2}{n_2} = \frac{390}{600} = 0.65, \ \overset{\mu}{p} = \frac{x_1 + x_2}{n_1 + n_2} = \frac{272 + 390}{400 + 600} = 0.662$$

$$\sigma^\mu_{\bar{p}_1 - \bar{p}_2} = \sqrt{\left(\overset{\mu}{p}\right)\left(1 - \overset{\mu}{p}\right)\left(\frac{1}{n_1} + \frac{1}{n_2}\right)} = \sqrt{(0.662)(1 - 0.662)\left(\frac{1}{400} + \frac{1}{600}\right)} = 0.0305$$

$$z = \frac{(\bar{p}_1 - \bar{p}_2) - (p_1 - p_2)_{H_0}}{\sigma^{\mu}_{\bar{p}_1 - \bar{p}_2}} = \frac{(0.68 - 0.65) - 0}{0.0305} = +0.98, \; z_c = +2.33$$

Because $z \le z_c$, we do not reject H_0 and conclude there is not enough evidence to support the claim that the proportion of home ownership in Florida is greater than the national proportion.

p-value = $P[z > +0.98] = 1 - P[z \le +0.98] = (1 - 0.8365) = 0.1635$

Chapter 18

1. H_0: The arrival process can be described by the expected distribution.

   H_1: The arrival process differs from the expected distribution.

Sample Size = 215 Customers

Day	Expected Percentage	Sample Size	Expected Frequency (E)	Observed Frequency (O)
Mon	10%	215	0.10(215) = 21.5	31
Tues	10%	215	0.10(215) = 21.5	18
Wed	15%	215	0.15(215) = 32.25	36
Thurs	15%	215	0.15(215) = 32.25	23
Fri	20%	215	0.20(215) = 43	47
Sat	30%	215	0.30(215) = 64.5	60
Total	100%		215	215

Day	O	E	$(O - E)$	$(O - E)^2$	$\dfrac{(O - E)^2}{E}$
Mon	31	21.50	9.50	90.25	4.20
Tues	18	21.50	−3.50	12.25	0.57
Wed	36	32.25	3.75	14.06	0.44
Thurs	23	32.25	−9.25	85.56	2.65
Fri	47	43.00	4.00	16.00	0.37
Sat	60	64.50	−4.50	20.25	0.31
Total					$\chi^2 = \sum \dfrac{(O - E)^2}{E} = 8.54$

For $\alpha = 0.05$ and $d.f. = k - 1 = 6 - 1 = 5$, $\chi_c^2 = 12.592$. Because $\chi_c^2 > \chi^2$, we do not reject H_0 and conclude that the arrival distribution is consistent with the expected distribution.

2. H_0: The process can be described with the Poisson distribution using $\lambda = 3$.

   H_1: The process differs from the Poisson distribution using $\lambda = 3$.

Sample Size = 380 Hits

Number of Hits Per Minute	Poisson Probabilities	Number of Hits	Expected Frequency
0	0.0498 ×	380 =	18.92
1	0.1494 ×	380 =	56.77
2	0.2240 ×	380 =	85.12
3	0.2240 ×	380 =	85.12
4	0.1680 ×	380 =	63.84
5	0.1008 ×	380 =	38.30
6	0.0504 ×	380 =	19.15
7 or more	0.0336 ×	380 =	12.77
Total	1.0000		380.00

Hits per Min	O	E	$(O - E)$	$(O - E)^2$	$\dfrac{(O-E)^2}{E}$
0	22	18.92	3.08	9.46	0.50
1	51	56.77	−5.77	33.32	0.59
2	72	85.12	−13.12	172.13	2.02
3	92	85.12	6.88	47.33	0.56
4	60	63.84	−3.84	14.75	0.23
5	44	38.30	5.70	32.44	0.84
6	25	19.15	5.85	34.19	1.79
7 or more	14	12.77	1.23	1.52	0.12
Total					$\chi^2 = \sum \dfrac{(O-E)^2}{E} = 6.65$

For $\alpha = 0.01$ and $d.f. = k - 1 = 8 - 1 = 7$, $\chi_c^2 = 18.475$. Because $\chi_c^2 > \chi^2$, we do not reject H_0 and conclude that the process is consistent with the Poisson distribution using $\lambda = 3$.

3. H_0: Grades are independent of reading time.

H_1: Grades are dependent of reading time.

Sample expected frequency calculations:

$$E_{1,1} = \frac{(265)(95)}{500} = 50.35 \quad E_{1,2} = \frac{(265)(128)}{500} = 67.84 \quad E_{1,3} = \frac{(265)(155)}{500} = 82.15$$

Row	Column	O	E	$(O-E)$	$(O-E)^2$	$\frac{(O-E)^2}{E}$
1	1	36	50.35	−14.35	205.92	4.09
1	2	75	67.84	7.16	51.27	0.76
1	3	81	82.15	−1.15	1.32	0.02
1	4	63	49.82	13.18	173.71	3.49
1	5	10	14.84	−4.84	23.43	1.58
2	1	27	26.60	0.40	0.16	0.01
2	2	28	35.84	−7.84	61.47	1.72
2	3	50	43.40	6.60	43.56	1.00
2	4	25	26.32	−1.32	1.74	0.07
2	5	10	7.84	2.16	4.67	0.60
3	1	32	18.05	13.95	194.60	10.78
3	2	25	24.32	0.68	0.46	0.02
3	3	24	29.45	−5.45	29.70	1.01
3	4	6	17.86	−11.96	140.66	7.88
3	5	8	5.32	2.68	7.18	1.35

$$\text{Total} = \chi^2 = \sum \frac{(O-E)^2}{E} = 34.38$$

For $\alpha = 0.05$ and $d.f. = (r-1)(c-1) = (3-1)(5-1) = 8$, $\chi_c^2 = 15.507$. Because $\chi^2 > \chi_c^2$, we reject H_0 and conclude that there is a relationship between grades and number of hours reading.

Chapter 19

1. H_0: $\mu_1 = \mu_2 = \mu_3$ $\bar{x}_1 = 22.12$ $\bar{x}_2 = 19.67$ $\bar{x}_3 = 18.94$

H_1: not all μ's are created equal $s_1^2 = 0.98$ $s_2^2 = 1.45$ $s_3^2 = 2.36$

$N = 17$ $n_1 = 6$ $n_2 = 6$ $n_3 = 6$

$$SSW = \sum_{i=1}^{k}(n_i - 1)s_i^2 = (6-1)0.98 + (6-1)1.45 + (5-1)2.36 = 21.59$$

$$\overline{\overline{x}} = \frac{\sum x}{N} = \frac{22.5 + 20.8 + 22.0 + 23.6 + \ldots + 18.0 + 21.1 + 19.8 + 18.6}{17} = 20.32$$

$$SSB = \sum_{i=1}^{k}n_i\left(\overline{x}_i - \overline{\overline{x}}\right)^2 = 6(22.12 - 20.32)^2 + 6(19.67 - 20.32)^2 + 5(18.94 - 20.32)^2 = 31.50$$

$$MSB = \frac{SSB}{k-1} = \frac{31.50}{3-1} = 15.75 \; , \; MSW = \frac{SSW}{N-k} = \frac{21.59}{17-3} = 1.54,$$

$$F = \frac{MSB}{MSW} = \frac{15.75}{1.54} = 10.23, \; F_c = F_{\alpha,k-1,N-k} = F_{.05,2,14} = 3.739$$

Because $F > F_c$, we reject H_0 and conclude that there is a difference between the sample means.

2. For $\overline{x}_1$ and $\overline{x}_2$, $F_S = \dfrac{\left(\overline{x}_a - \overline{x}_b\right)^2}{\dfrac{SSW}{\sum\limits_{i=1}^{k}(n_i - 1)}\left[\dfrac{1}{n_a} + \dfrac{1}{n_b}\right]} = \dfrac{(22.12 - 19.67)^2}{\dfrac{21.59}{5+5+4}\left[\dfrac{1}{6} + \dfrac{1}{6}\right]} = 11.70$

For $\overline{x}_1$ and $\overline{x}_3$, $F_S = \dfrac{\left(\overline{x}_a - \overline{x}_b\right)^2}{\dfrac{SSW}{\sum\limits_{i=1}^{k}(n_i - 1)}\left[\dfrac{1}{n_a} + \dfrac{1}{n_b}\right]} = \dfrac{(22.12 - 18.94)^2}{\dfrac{21.59}{5+5+4}\left[\dfrac{1}{6} + \dfrac{1}{6}\right]} = 19.71$

For $\overline{x}_2$ and $\overline{x}_3$, $F_S = \dfrac{\left(\overline{x}_a - \overline{x}_b\right)^2}{\dfrac{SSW}{\sum\limits_{i=1}^{k}(n_i - 1)}\left[\dfrac{1}{n_a} + \dfrac{1}{n_b}\right]} = \dfrac{(19.67 - 18.94)^2}{\dfrac{21.59}{5+5+4}\left[\dfrac{1}{6} + \dfrac{1}{6}\right]} = 1.04$

$$F_{SC} = (k-1)F_{\alpha,k-1,N-k} = (3-1)(3.739) = 7.478$$

Sample Pair	F_S	F_{SC}	Conclusion
$\overline{x}_1$ and $\overline{x}_2$	11.70	7.478	Difference
$\overline{x}_1$ and $\overline{x}_3$	19.71	7.478	Difference
$\overline{x}_2$ and $\overline{x}_3$	1.04	7.478	No Difference

We conclude that there is a difference between gas mileage of Cars 1 and 2 and a difference between Cars 1 and 3.

3. H_0: $\mu_1 = \mu_2 = \mu_3$ $\bar{x}_1 = 38.33$ $\bar{x}_2 = 28.29$ $\bar{x}_3 = 28.0$ $\bar{x}_4 = 31.43$

H_1: not all μ's are created equal $s_1^2 = 115.47$ $s_2^2 = 72.57$ $s_3^2 = 86.8$ $s_4^2 = 132.62$

$N = 26$ $n_1 = 6$ $n_2 = 7$ $n_3 = 6$ $n_4 = 7$

$SSW = (6-1)115.47 + (7-1)72.57 + (6-1)86.8 + (7-1)132.62 = 2242.49$

$$\bar{\bar{x}} = \frac{\sum x}{N} = \frac{36+48+32+28+...+36+18+30+21}{26} = 31.38$$

$SSB = 6(38.33-31.38)^2 + 7(28.29-31.38)^2 + 6(28-31.38)^2 + 7(31.43-31.38)^2 = 425.22$

$$MSB = \frac{SSB}{k-1} = \frac{425.22}{4-1} = 141.74, \quad MSW = \frac{SSW}{N-k} = \frac{2242.49}{26-4} = 101.93$$

$$F = \frac{MSB}{MSW} = \frac{141.74}{101.93} = 1.391, \quad F_c = F_{\alpha,k-1,N-k} = F_{.05,3,22} = 3.049$$

Because $F \le F_c$, we do not reject H_0 and conclude that there is no difference between the sample means.

Chapter 20

1.

Payroll x	Wins y	xy	x^2	y^2
171	103	17613	29241	10609
108	75	8100	11664	5625
119	92	10948	14161	8464
43	55	2365	1849	3025
58	56	3248	3364	3136
56	62	3472	3136	3844
62	84	5208	3844	7056
43	78	3354	1849	6084
57	73	4161	3249	5329
75	67	5025	5625	4489
$\sum x = 745$	$\sum y = 792$	$\sum xy = 63,494$	$\sum x^2 = 77,982$	$\sum y^2 = 57,661$

$$\bar{x} = \frac{792}{10} = 79.2, \quad \bar{y} = \frac{745}{10} = 74.5, \quad r = \frac{n\sum xy - \left(\sum x\right)\left(\sum y\right)}{\sqrt{\left[n\sum x^2 - \left(\sum x\right)^2\right]\left[n\sum y^2 - \left(\sum y\right)^2\right]}}$$

$$r = \frac{10(63,494) - (792)(745)}{\sqrt{\left[10(77,982) - (792)^2\right]\left[10(57,661) - (745)^2\right]}} = \frac{44,900}{\sqrt{(152,556)(21,585)}} = 0.782$$

$H_0 : \rho = 0$

$H_1 : \rho \neq 0$

$$t = \frac{r}{\sqrt{\dfrac{1-r^2}{n-2}}} = \frac{0.782}{\sqrt{\dfrac{1-(0.782)^2}{10-2}}} = 3.549 \text{ , } d.f. = n - 2 = 10 - 2 = 8, \ t_c = 2.306$$

Because $t > t_c$, we reject H_0 and conclude the correlation coefficient is not equal to 0.

2a. $\displaystyle b = \frac{n\sum xy - \left(\sum x\right)\left(\sum y\right)}{n\sum x^2 - \left(\sum x\right)^2} = \frac{10(63,494) - (792)(745)}{10(77,982) - (792)^2} = \frac{44,900}{152,556} = 0.294$

$a = \bar{y} - b\bar{x} = 74.5 - (0.294)79.2 = 51.21 \text{ , } \hat{y} = 51.21 + 0.294x$

2b. $H_0 : \beta = 0$

$H_1 : \beta \neq 0$

$$s_e = \sqrt{\frac{\sum y^2 - a\sum y - b\sum xy}{n-2}} = \sqrt{\frac{(57,661) - 51.21(745) - (0.294)(63,494)}{10-2}} = 10.26$$

$$s_b = \frac{s_e}{\sqrt{\sum x^2 - n\bar{x}^2}} = \frac{10.26}{\sqrt{77,982 - 10(79.2)^2}} = 0.0831, \ t = \frac{b - \beta_{H_0}}{s_b} = \frac{0.294 - 0}{0.0831} = 3.538$$

$d.f. = n - 2 = 10 - 2 = 8$, $t_c = 2.306$, because $t > t_c$, we reject H_0 and conclude there is a relationship between payroll and wins.

2c. $\hat{y} = 51.21 + 0.294x = 51.21 + 0.294(70) = 71.79$

2d. $\displaystyle CI = \hat{y} \pm t_c s_e \sqrt{\frac{1}{n} + \frac{(x - \bar{x})^2}{\left(\sum x^2\right) - \dfrac{\left(\sum x\right)^2}{n}}}$

$d.f. = n - 2 = 10 - 2 = 8$, $t_c = 3.355$

$$CI = 71.79 \pm (3.355)(10.26)\sqrt{\frac{1}{10} + \frac{(70 - 79.2)^2}{(77,982) - \dfrac{(792)^2}{10}}}$$

$CI = 71.79 \pm (3.355)(10.26)(0.325) = 71.79 \pm 11.19, \ (60.60, 82.98)$

2e. $r^2 = (0.782)^2 = 0.612$ or 61.2 percent

Statistical Tables

Source: Mr. Carl Schwarz, www.stat.sfu.ca/~cschwarz. Used with permission.

Table 1 provides the probability of exactly r successes in n trials for various values of p.

Table 1 Binomial Probability Tables

		Values of p								
n	r	0.1	0.2	0.3	0.4	0.5	0.6	0.7	0.8	0.9
2	0	0.8100	0.6400	0.4900	0.3600	0.2500	0.1600	0.0900	0.0400	0.0100
	1	0.1800	0.3200	0.4200	0.4800	0.5000	0.4800	0.4200	0.3200	0.1800
	2	0.0100	0.0400	0.0900	0.1600	0.2500	0.3600	0.4900	0.6400	0.8100
3	0	0.7290	0.5120	0.3430	0.2160	0.1250	0.0640	0.0270	0.0080	0.0010
	1	0.2430	0.3840	0.4410	0.4320	0.3750	0.2880	0.1890	0.0960	0.0270
	2	0.0270	0.0960	0.1890	0.2880	0.3750	0.4320	0.4410	0.3840	0.2430
	3	0.0010	0.0080	0.0270	0.0640	0.1250	0.2160	0.3430	0.5120	0.7290
4	0	0.6561	0.4096	0.2401	0.1296	0.0625	0.0256	0.0081	0.0016	0.0001
	1	0.2916	0.4096	0.4116	0.3456	0.2500	0.1536	0.0756	0.0256	0.0036
	2	0.0486	0.1536	0.2646	0.3456	0.3750	0.3456	0.2646	0.1536	0.0486
	3	0.0036	0.0256	0.0756	0.1536	0.2500	0.3456	0.4116	0.4096	0.2916
	4	0.0001	0.0016	0.0081	0.0256	0.0625	0.1296	0.2401	0.4096	0.6561

Table 1   Binomial Probability Tables   (continued)

Values of p

n	r	0.1	0.2	0.3	0.4	0.5	0.6	0.7	0.8	0.9
5	0	0.5905	0.3277	0.1681	0.0778	0.0313	0.0102	0.0024	0.0003	0.0000
	1	0.3280	0.4096	0.3601	0.2592	0.1563	0.0768	0.0284	0.0064	0.0005
	2	0.0729	0.2048	0.3087	0.3456	0.3125	0.2304	0.1323	0.0512	0.0081
	3	0.0081	0.0512	0.1323	0.2304	0.3125	0.3456	0.3087	0.2048	0.0729
	4	0.0005	0.0064	0.0283	0.0768	0.1563	0.2592	0.3601	0.4096	0.3281
	5	0.0000	0.0003	0.0024	0.0102	0.0313	0.0778	0.1681	0.3277	0.5905
6	0	0.5314	0.2621	0.1176	0.0467	0.0156	0.0041	0.0007	0.0001	0.0000
	1	0.3543	0.3932	0.3025	0.1866	0.0938	0.0369	0.0102	0.0015	0.0001
	2	0.0984	0.2458	0.3241	0.3110	0.2344	0.1382	0.0595	0.0154	0.0012
	3	0.0146	0.0819	0.1852	0.2765	0.3125	0.2765	0.1852	0.0819	0.0146
	4	0.0012	0.0154	0.0595	0.1382	0.2344	0.3110	0.3241	0.2458	0.0984
	5	0.0001	0.0015	0.0102	0.0369	0.0938	0.1866	0.3025	0.3932	0.3543
	6	0.0000	0.0001	0.0007	0.0041	0.0156	0.0467	0.1176	0.2621	0.5314
7	0	0.4783	0.2097	0.0824	0.0280	0.0078	0.0016	0.0002	0.0000	0.0000
	1	0.3720	0.3670	0.2471	0.1306	0.0547	0.0172	0.0036	0.0004	0.0000
	2	0.1240	0.2753	0.3177	0.2613	0.1641	0.0774	0.0250	0.0043	0.0002
	3	0.0230	0.1147	0.2269	0.2903	0.2734	0.1935	0.0972	0.0287	0.0026
	4	0.0026	0.0287	0.0972	0.1935	0.2734	0.2903	0.2269	0.1147	0.0230
	5	0.0002	0.0043	0.0250	0.0774	0.1641	0.2613	0.3177	0.2753	0.1240
	6	0.0000	0.0004	0.0036	0.0172	0.0547	0.1306	0.2471	0.3670	0.3720
	7	0.0000	0.0000	0.0002	0.0016	0.0078	0.0280	0.0824	0.2097	0.4783
8	0	0.4305	0.1678	0.0576	0.0168	0.0039	0.0007	0.0001	0.0000	0.0000
	1	0.3826	0.3355	0.1977	0.0896	0.0313	0.0079	0.0012	0.0001	0.0000
	2	0.1488	0.2936	0.2965	0.2090	0.1094	0.0413	0.0100	0.0011	0.0000
	3	0.0331	0.1468	0.2541	0.2787	0.2188	0.1239	0.0467	0.0092	0.0004
	4	0.0046	0.0459	0.1361	0.2322	0.2734	0.2322	0.1361	0.0459	0.0046
	5	0.0004	0.0092	0.0467	0.1239	0.2188	0.2787	0.2541	0.1468	0.0331
	6	0.0000	0.0011	0.0100	0.0413	0.1094	0.2090	0.2965	0.2936	0.1488
	7	0.0000	0.0001	0.0012	0.0079	0.0313	0.0896	0.1977	0.3355	0.3826
	8	0.0000	0.0000	0.0001	0.0007	0.0039	0.0168	0.0576	0.1678	0.4305

Values of p

n	r	0.1	0.2	0.3	0.4	0.5	0.6	0.7	0.8	0.9
9	0	0.3874	0.1342	0.0404	0.0101	0.0020	0.0003	0.0000	0.0000	0.0000
	1	0.3874	0.3020	0.1556	0.0605	0.0176	0.0035	0.0004	0.0000	0.0000
	2	0.1722	0.3020	0.2668	0.1612	0.0703	0.0212	0.0039	0.0003	0.0000
	3	0.0446	0.1762	0.2668	0.2508	0.1641	0.0743	0.0210	0.0028	0.0001
	4	0.0074	0.0661	0.1715	0.2508	0.2461	0.1672	0.0735	0.0165	0.0008
	5	0.0008	0.0165	0.0735	0.1672	0.2461	0.2508	0.1715	0.0661	0.0074
	6	0.0001	0.0028	0.0210	0.0743	0.1641	0.2508	0.2668	0.1762	0.0446
	7	0.0000	0.0003	0.0039	0.0212	0.0703	0.1612	0.2668	0.3020	0.1722
	8	0.0000	0.0000	0.0004	0.0035	0.0176	0.0605	0.1556	0.3020	0.3874
	9	0.0000	0.0000	0.0000	0.0003	0.0020	0.0101	0.0404	0.1342	0.3874
10	0	0.3487	0.1074	0.0282	0.0060	0.0010	0.0001	0.0000	0.0000	0.0000
	1	0.3874	0.2684	0.1211	0.0403	0.0098	0.0016	0.0001	0.0000	0.0000
	2	0.1937	0.3020	0.2335	0.1209	0.0439	0.0106	0.0014	0.0001	0.0000
	3	0.0574	0.2013	0.2668	0.2150	0.1172	0.0425	0.0090	0.0008	0.0000
	4	0.0112	0.0881	0.2001	0.2508	0.2051	0.1115	0.0368	0.0055	0.0001
	5	0.0015	0.0264	0.1029	0.2007	0.2461	0.2007	0.1029	0.0264	0.0015
	6	0.0001	0.0055	0.0368	0.1115	0.2051	0.2508	0.2001	0.0881	0.0112
	7	0.0000	0.0008	0.0090	0.0425	0.1172	0.2150	0.2668	0.2013	0.0574
	8	0.0000	0.0001	0.0014	0.0106	0.0439	0.1209	0.2335	0.3020	0.1937
	9	0.0000	0.0000	0.0001	0.0016	0.0098	0.0403	0.1211	0.2684	0.3874
	10	0.0000	0.0000	0.0000	0.0001	0.0010	0.0060	0.0282	0.1074	0.3487
11	0	0.3138	0.0859	0.0198	0.0036	0.0005	0.0000	0.0000	0.0000	0.0000
	1	0.3835	0.2362	0.0932	0.0266	0.0054	0.0007	0.0000	0.0000	0.0000
	2	0.2131	0.2953	0.1998	0.0887	0.0269	0.0052	0.0005	0.0000	0.0000
	3	0.0710	0.2215	0.2568	0.1774	0.0806	0.0234	0.0037	0.0002	0.0000
	4	0.0158	0.1107	0.2201	0.2365	0.1611	0.0701	0.0173	0.0017	0.0000
	5	0.0025	0.0388	0.1321	0.2207	0.2256	0.1471	0.0566	0.0097	0.0003
	6	0.0003	0.0097	0.0566	0.1471	0.2256	0.2207	0.1321	0.0388	0.0025
	7	0.0000	0.0017	0.0173	0.0701	0.1611	0.2365	0.2201	0.1107	0.0158
	8	0.0000	0.0002	0.0037	0.0234	0.0806	0.1774	0.2568	0.2215	0.0710
	9	0.0000	0.0000	0.0005	0.0052	0.0269	0.0887	0.1998	0.2953	0.2131

continues

Table 1 Binomial Probability Tables (continued)

Values of p

n	r	0.1	0.2	0.3	0.4	0.5	0.6	0.7	0.8	0.9
	10	0.0000	0.0000	0.0000	0.0007	0.0054	0.0266	0.0932	0.2362	0.3835
	11	0.0000	0.0000	0.0000	0.0000	0.0005	0.0036	0.0198	0.0859	0.3138
12	0	0.2824	0.0687	0.0138	0.0022	0.0002	0.0000	0.0000	0.0000	0.0000
	1	0.3766	0.2062	0.0712	0.0174	0.0029	0.0003	0.0000	0.0000	0.0000
	2	0.2301	0.2835	0.1678	0.0639	0.0161	0.0025	0.0002	0.0000	0.0000
	3	0.0852	0.2362	0.2397	0.1419	0.0537	0.0125	0.0015	0.0001	0.0000
	4	0.0213	0.1329	0.2311	0.2128	0.1208	0.0420	0.0078	0.0005	0.0000
	5	0.0038	0.0532	0.1585	0.2270	0.1934	0.1009	0.0291	0.0033	0.0000
	6	0.0005	0.0155	0.0792	0.1766	0.2256	0.1766	0.0792	0.0155	0.0005
	7	0.0000	0.0033	0.0291	0.1009	0.1934	0.2270	0.1585	0.0532	0.0038
	8	0.0000	0.0005	0.0078	0.0420	0.1208	0.2128	0.2311	0.1329	0.0213
	9	0.0000	0.0001	0.0015	0.0125	0.0537	0.1419	0.2397	0.2362	0.0852
	10	0.0000	0.0000	0.0002	0.0025	0.0161	0.0639	0.1678	0.2835	0.2301
	11	0.0000	0.0000	0.0000	0.0003	0.0029	0.0174	0.0712	0.2062	0.3766
	12	0.0000	0.0000	0.0000	0.0000	0.0002	0.0022	0.0138	0.0687	0.2824
13	0	0.2542	0.0550	0.0097	0.0013	0.0001	0.0000	0.0000	0.0000	0.0000
	1	0.3672	0.1787	0.0540	0.0113	0.0016	0.0001	0.0000	0.0000	0.0000
	2	0.2448	0.2680	0.1388	0.0453	0.0095	0.0012	0.0001	0.0000	0.0000
	3	0.0997	0.2457	0.2181	0.1107	0.0349	0.0065	0.0006	0.0000	0.0000
	4	0.0277	0.1535	0.2337	0.1845	0.0873	0.0243	0.0034	0.0001	0.0000
	5	0.0055	0.0691	0.1803	0.2214	0.1571	0.0656	0.0142	0.0011	0.0000
	6	0.0008	0.0230	0.1030	0.1968	0.2095	0.1312	0.0442	0.0058	0.0001
	7	0.0001	0.0058	0.0442	0.1312	0.2095	0.1968	0.1030	0.0230	0.0008
	8	0.0000	0.0011	0.0142	0.0656	0.1571	0.2214	0.1803	0.0691	0.0055
	9	0.0000	0.0001	0.0034	0.0243	0.0873	0.1845	0.2337	0.1535	0.0277
	10	0.0000	0.0000	0.0006	0.0065	0.0349	0.1107	0.2181	0.2457	0.0997
	11	0.0000	0.0000	0.0001	0.0012	0.0095	0.0453	0.1388	0.2680	0.2448
	12	0.0000	0.0000	0.0000	0.0001	0.0016	0.0113	0.0540	0.1787	0.3672
	13	0.0000	0.0000	0.0000	0.0000	0.0001	0.0013	0.0097	0.0550	0.2542

Values of p

n	r	0.1	0.2	0.3	0.4	0.5	0.6	0.7	0.8	0.9
14	0	0.2288	0.0440	0.0068	0.0008	0.0001	0.0000	0.0000	0.0000	0.0000
	1	0.3559	0.1539	0.0407	0.0073	0.0009	0.0001	0.0000	0.0000	0.0000
	2	0.2570	0.2501	0.1134	0.0317	0.0056	0.0005	0.0000	0.0000	0.0000
	3	0.1142	0.2501	0.1943	0.0845	0.0222	0.0033	0.0002	0.0000	0.0000
	4	0.0349	0.1720	0.2290	0.1549	0.0611	0.0136	0.0014	0.0000	0.0000
	5	0.0078	0.0860	0.1963	0.2066	0.1222	0.0408	0.0066	0.0003	0.0000
	6	0.0013	0.0322	0.1262	0.2066	0.1833	0.0918	0.0232	0.0020	0.0000
	7	0.0002	0.0092	0.0618	0.1574	0.2095	0.1574	0.0618	0.0092	0.0002
	8	0.0000	0.0020	0.0232	0.0918	0.1833	0.2066	0.1262	0.0322	0.0013
	9	0.0000	0.0003	0.0066	0.0408	0.1222	0.2066	0.1963	0.0860	0.0078
	10	0.0000	0.0000	0.0014	0.0136	0.0611	0.1549	0.2290	0.1720	0.0349
	11	0.0000	0.0000	0.0002	0.0033	0.0222	0.0845	0.1943	0.2501	0.1142
	12	0.0000	0.0000	0.0000	0.0005	0.0056	0.0317	0.1134	0.2501	0.2570
	13	0.0000	0.0000	0.0000	0.0001	0.0009	0.0073	0.0407	0.1539	0.3559
	14	0.0000	0.0000	0.0000	0.0000	0.0001	0.0008	0.0068	0.0440	0.2288
15	0	0.2059	0.0352	0.0047	0.0005	0.0000	0.0000	0.0000	0.0000	0.0000
	1	0.3432	0.1319	0.0305	0.0047	0.0005	0.0000	0.0000	0.0000	0.0000
	2	0.2669	0.2309	0.0916	0.0219	0.0032	0.0003	0.0000	0.0000	0.0000
	3	0.1285	0.2501	0.1700	0.0634	0.0139	0.0016	0.0001	0.0000	0.0000
	4	0.0428	0.1876	0.2186	0.1268	0.0417	0.0074	0.0006	0.0000	0.0000
	5	0.0105	0.1032	0.2061	0.1859	0.0916	0.0245	0.0030	0.0001	0.0000
	6	0.0019	0.0430	0.1472	0.2066	0.1527	0.0612	0.0116	0.0007	0.0000
	7	0.0003	0.0138	0.0811	0.1771	0.1964	0.1181	0.0348	0.0035	0.0000
	8	0.0000	0.0035	0.0348	0.1181	0.1964	0.1771	0.0811	0.0138	0.0003
	9	0.0000	0.0007	0.0116	0.0612	0.1527	0.2066	0.1472	0.0430	0.0019
	10	0.0000	0.0001	0.0030	0.0245	0.0916	0.1859	0.2061	0.1032	0.0105
	11	0.0000	0.0000	0.0006	0.0074	0.0417	0.1268	0.2186	0.1876	0.0428
	12	0.0000	0.0000	0.0001	0.0016	0.0139	0.0634	0.1700	0.2501	0.1285
	13	0.0000	0.0000	0.0000	0.0003	0.0032	0.0219	0.0916	0.2309	0.2669
	14	0.0000	0.0000	0.0000	0.0000	0.0005	0.0047	0.0305	0.1319	0.3432
	15	0.0000	0.0000	0.0000	0.0000	0.0000	0.0005	0.0047	0.0352	0.2059

Table 2 provides the probability of exactly *x* number of occurrences for various values of μ.

Table 2   Poisson Probability Tables

Values of μ

x	0.1	0.2	0.3	0.4	0.5	0.6	0.7	0.8	0.9	1.0
0	0.9048	0.8187	0.7408	0.6703	0.6065	0.5488	0.4966	0.4493	0.4066	0.3679
1	0.0905	0.1637	0.2222	0.2681	0.3033	0.3293	0.3476	0.3595	0.3659	0.3679
2	0.0045	0.0164	0.0333	0.0536	0.0758	0.0988	0.1217	0.1438	0.1647	0.1839
3	0.0002	0.0011	0.0033	0.0072	0.0126	0.0198	0.0284	0.0383	0.0494	0.0613
4	0.0000	0.0001	0.0003	0.0007	0.0016	0.0030	0.0050	0.0077	0.0111	0.0153
5	0.0000	0.0000	0.0000	0.0001	0.0002	0.0004	0.0007	0.0012	0.0020	0.0031
6	0.0000	0.0000	0.0000	0.0000	0.0000	0.0000	0.0001	0.0002	0.0003	0.0005

Values of μ

x	1.1	1.2	1.3	1.4	1.5	1.6	1.7	1.8	1.9	2.0
0	0.3329	0.3012	0.2725	0.2466	0.2231	0.2019	0.1827	0.1653	0.1496	0.1353
1	0.3662	0.3614	0.3543	0.3452	0.3347	0.3230	0.3106	0.2975	0.2842	0.2707
2	0.2014	0.2169	0.2303	0.2417	0.2510	0.2584	0.2640	0.2678	0.2700	0.2707
3	0.0738	0.0867	0.0998	0.1128	0.1255	0.1378	0.1496	0.1607	0.1710	0.1804
4	0.0203	0.0260	0.0324	0.0395	0.0471	0.0551	0.0636	0.0723	0.0812	0.0902
5	0.0045	0.0062	0.0084	0.0111	0.0141	0.0176	0.0216	0.0260	0.0309	0.0361
6	0.0008	0.0012	0.0018	0.0026	0.0035	0.0047	0.0061	0.0078	0.0098	0.0120
7	0.0001	0.0002	0.0003	0.0005	0.0008	0.0011	0.0015	0.0020	0.0027	0.0034
8	0.0000	0.0000	0.0001	0.0001	0.0001	0.0002	0.0003	0.0005	0.0006	0.0009
9	0.0000	0.0000	0.0000	0.0000	0.0000	0.0000	0.0001	0.0001	0.0001	0.0002

Values of μ

x	2.1	2.2	2.3	2.4	2.5	2.6	2.7	2.8	2.9	3.0
0	0.1225	0.1108	0.1003	0.0907	0.0821	0.0743	0.0672	0.0608	0.0550	0.0498
1	0.2572	0.2438	0.2306	0.2177	0.2052	0.1931	0.1815	0.1703	0.1596	0.1494
2	0.2700	0.2681	0.2652	0.2613	0.2565	0.2510	0.2450	0.2384	0.2314	0.2240
3	0.1890	0.1966	0.2033	0.2090	0.2138	0.2176	0.2205	0.2225	0.2237	0.2240
4	0.0992	0.1082	0.1169	0.1254	0.1336	0.1414	0.1488	0.1557	0.1622	0.1680
5	0.0417	0.0476	0.0538	0.0602	0.0668	0.0735	0.0804	0.0872	0.0940	0.1008
6	0.0146	0.0174	0.0206	0.0241	0.0278	0.0319	0.0362	0.0407	0.0455	0.0504

Values of μ

7	0.0044	0.0055	0.0068	0.0083	0.0099	0.0118	0.0139	0.0163	0.0188	0.0216
8	0.0011	0.0015	0.0019	0.0025	0.0031	0.0038	0.0047	0.0057	0.0068	0.0081
9	0.0003	0.0004	0.0005	0.0007	0.0009	0.0011	0.0014	0.0018	0.0022	0.0027
10	0.0001	0.0001	0.0001	0.0002	0.0002	0.0003	0.0004	0.0005	0.0006	0.0008
11	0.0000	0.0000	0.0000	0.0000	0.0000	0.0001	0.0001	0.0001	0.0002	0.0002

Values of μ

x	3.2	3.4	3.6	3.8	4.0	4.2	4.4	4.6	4.8	5.0
0	0.0408	0.0334	0.0273	0.0224	0.0183	0.0150	0.0123	0.0101	0.0082	0.0067
1	0.1304	0.1135	0.0984	0.0850	0.0733	0.0630	0.0540	0.0462	0.0395	0.0337
2	0.2087	0.1929	0.1771	0.1615	0.1465	0.1323	0.1188	0.1063	0.0948	0.0842
3	0.2226	0.2186	0.2125	0.2046	0.1954	0.1852	0.1743	0.1631	0.1517	0.1404
4	0.1781	0.1858	0.1912	0.1944	0.1954	0.1944	0.1917	0.1875	0.1820	0.1755
5	0.1140	0.1264	0.1377	0.1477	0.1563	0.1633	0.1687	0.1725	0.1747	0.1755
6	0.0608	0.0716	0.0826	0.0936	0.1042	0.1143	0.1237	0.1323	0.1398	0.1462
7	0.0278	0.0348	0.0425	0.0508	0.0595	0.0686	0.0778	0.0869	0.0959	0.1044
8	0.0111	0.0148	0.0191	0.0241	0.0298	0.0360	0.0428	0.0500	0.0575	0.0653
9	0.0040	0.0056	0.0076	0.0102	0.0132	0.0168	0.0209	0.0255	0.0307	0.0363
10	0.0013	0.0019	0.0028	0.0039	0.0053	0.0071	0.0092	0.0118	0.0147	0.0181
11	0.0004	0.0006	0.0009	0.0013	0.0019	0.0027	0.0037	0.0049	0.0064	0.0082
12	0.0001	0.0002	0.0003	0.0004	0.0006	0.0009	0.0013	0.0019	0.0026	0.0034
13	0.0000	0.0000	0.0001	0.0001	0.0002	0.0003	0.0005	0.0007	0.0009	0.0013
14	0.0000	0.0000	0.0000	0.0000	0.0001	0.0001	0.0001	0.0002	0.0003	0.0005
15	0.0000	0.0000	0.0000	0.0000	0.0000	0.0000	0.0000	0.0001	0.0001	0.0002

Values of μ

x	5.2	5.4	5.6	5.8	6.0	6.2	6.4	6.6	6.8	7.0
0	0.0055	0.0045	0.0037	0.0030	0.0025	0.0020	0.0017	0.0014	0.0011	0.0009
1	0.0287	0.0244	0.0207	0.0176	0.0149	0.0126	0.0106	0.0090	0.0076	0.0064
2	0.0746	0.0659	0.0580	0.0509	0.0446	0.0390	0.0340	0.0296	0.0258	0.0223
3	0.1293	0.1185	0.1082	0.0985	0.0892	0.0806	0.0726	0.0652	0.0584	0.0521
4	0.1681	0.1600	0.1515	0.1428	0.1339	0.1249	0.1162	0.1076	0.0992	0.0912
5	0.1748	0.1728	0.1697	0.1656	0.1606	0.1549	0.1487	0.1420	0.1349	0.1277
6	0.1515	0.1555	0.1584	0.1601	0.1606	0.1601	0.1586	0.1562	0.1529	0.1490

continues

Table 2 Poisson Probability Tables   (continued)

Values of μ										
7	0.1125	0.1200	0.1267	0.1326	0.1377	0.1418	0.1450	0.1472	0.1486	0.1490
8	0.0731	0.0810	0.0887	0.0962	0.1033	0.1099	0.1160	0.1215	0.1263	0.1304
9	0.0423	0.0486	0.0552	0.0620	0.0688	0.0757	0.0825	0.0891	0.0954	0.1014
10	0.0220	0.0262	0.0309	0.0359	0.0413	0.0469	0.0528	0.0588	0.0649	0.0710
11	0.0104	0.0129	0.0157	0.0190	0.0225	0.0265	0.0307	0.0353	0.0401	0.0452
12	0.0045	0.0058	0.0073	0.0092	0.0113	0.0137	0.0164	0.0194	0.0227	0.0263
13	0.0018	0.0024	0.0032	0.0041	0.0052	0.0065	0.0081	0.0099	0.0119	0.0142
14	0.0007	0.0009	0.0013	0.0017	0.0022	0.0029	0.0037	0.0046	0.0058	0.0071
15	0.0002	0.0003	0.0005	0.0007	0.0009	0.0012	0.0016	0.0020	0.0026	0.0033
16	0.0001	0.0001	0.0002	0.0002	0.0003	0.0005	0.0006	0.0008	0.0011	0.0014
17	0.0000	0.0000	0.0001	0.0001	0.0001	0.0002	0.0002	0.0003	0.0004	0.0006
18	0.0000	0.0000	0.0000	0.0000	0.0000	0.0001	0.0001	0.0001	0.0002	0.0002

Table 3 provides the area to the left of the corresponding z-score for the standard normal distribution.

Table 3 Normal Probability Tables

Second digit of z										
z	*0.00*	*0.01*	*0.02*	*0.03*	*0.04*	*0.05*	*0.06*	*0.07*	*0.08*	*0.09*
0.0	0.5000	0.5040	0.5080	0.5120	0.5160	0.5199	0.5239	0.5279	0.5319	0.5359
0.1	0.5398	0.5438	0.5478	0.5517	0.5557	0.5596	0.5636	0.5675	0.5714	0.5753
0.2	0.5793	0.5832	0.5871	0.5910	0.5948	0.5987	0.6026	0.6064	0.6103	0.6141
0.3	0.6179	0.6217	0.6255	0.6293	0.6331	0.6368	0.6406	0.6443	0.6480	0.6517
0.4	0.6554	0.6591	0.6628	0.6664	0.6700	0.6736	0.6772	0.6808	0.6844	0.6879
0.5	0.6915	0.6950	0.6985	0.7019	0.7054	0.7088	0.7123	0.7157	0.7190	0.7224
0.6	0.7257	0.7291	0.7324	0.7357	0.7389	0.7422	0.7454	0.7486	0.7517	0.7549
0.7	0.7580	0.7611	0.7642	0.7673	0.7704	0.7734	0.7764	0.7794	0.7823	0.7852
0.8	0.7881	0.7910	0.7939	0.7967	0.7995	0.8023	0.8051	0.8078	0.8106	0.8133
0.9	0.8159	0.8186	0.8212	0.8238	0.8264	0.8289	0.8315	0.8340	0.8365	0.8389

Second digit of z

z	0.00	0.01	0.02	0.03	0.04	0.05	0.06	0.07	0.08	0.09
1.0	0.8413	0.8438	0.8461	0.8485	0.8508	0.8531	0.8554	0.8577	0.8599	0.8621
1.1	0.8643	0.8665	0.8686	0.8708	0.8729	0.8749	0.8770	0.8790	0.8810	0.8830
1.2	0.8849	0.8869	0.8888	0.8907	0.8925	0.8944	0.8962	0.8980	0.8997	0.9015
1.3	0.9032	0.9049	0.9066	0.9082	0.9099	0.9115	0.9131	0.9147	0.9162	0.9177
1.4	0.9192	0.9207	0.9222	0.9236	0.9251	0.9265	0.9279	0.9292	0.9306	0.9319
1.5	0.9332	0.9345	0.9357	0.9370	0.9382	0.9394	0.9406	0.9418	0.9429	0.9441
1.6	0.9452	0.9463	0.9474	0.9484	0.9495	0.9505	0.9515	0.9525	0.9535	0.9545
1.7	0.9554	0.9564	0.9573	0.9582	0.9591	0.9599	0.9608	0.9616	0.9625	0.9633
1.8	0.9641	0.9649	0.9656	0.9664	0.9671	0.9678	0.9686	0.9693	0.9699	0.9706
1.9	0.9713	0.9719	0.9726	0.9732	0.9738	0.9744	0.9750	0.9756	0.9761	0.9767
2.0	0.9772	0.9778	0.9783	0.9788	0.9793	0.9798	0.9803	0.9808	0.9812	0.9817
2.1	0.9821	0.9826	0.9830	0.9834	0.9838	0.9842	0.9846	0.9850	0.9854	0.9857
2.2	0.9861	0.9864	0.9868	0.9871	0.9875	0.9878	0.9881	0.9884	0.9887	0.9890
2.3	0.9893	0.9896	0.9898	0.9901	0.9904	0.9906	0.9909	0.9911	0.9913	0.9916
2.4	0.9918	0.9920	0.9922	0.9925	0.9927	0.9929	0.9931	0.9932	0.9934	0.9936
2.5	0.9938	0.9940	0.9941	0.9943	0.9945	0.9946	0.9948	0.9949	0.9951	0.9952
2.6	0.9953	0.9955	0.9956	0.9957	0.9959	0.9960	0.9961	0.9962	0.9963	0.9964
2.7	0.9965	0.9966	0.9967	0.9968	0.9969	0.9970	0.9971	0.9972	0.9973	0.9974
2.8	0.9974	0.9975	0.9976	0.9977	0.9977	0.9978	0.9979	0.9979	0.9980	0.9981
2.9	0.9981	0.9982	0.9982	0.9983	0.9984	0.9984	0.9985	0.9985	0.9986	0.9986
3.0	0.9987	0.9987	0.9987	0.9988	0.9988	0.9989	0.9989	0.9989	0.9990	0.9990
3.1	0.9990	0.9991	0.9991	0.9991	0.9992	0.9992	0.9992	0.9992	0.9993	0.9993
3.2	0.9993	0.9993	0.9994	0.9994	0.9994	0.9994	0.9994	0.9995	0.9995	0.9995
3.3	0.9995	0.9995	0.9995	0.9996	0.9996	0.9996	0.9996	0.9996	0.9996	0.9997
3.4	0.9997	0.9997	0.9997	0.9997	0.9997	0.9997	0.9997	0.9997	0.9997	0.9998
3.5	0.9998	0.9998	0.9998	0.9998	0.9998	0.9998	0.9998	0.9998	0.9998	0.9998

Table 4 provides the t-statistic for the corresponding value of alpha or confidence interval and the number of degrees of freedom.

Table 4   Student's t-Distribution

Selected right-tail areas with confidence levels underneath

conf lev df	0.2000 0.6000	0.1500 0.7000	0.1000 0.8000	0.0500 0.9000	0.0250 0.9500	0.0100 0.9800	0.0050 0.9900	0.0010 0.9980	0.0005 0.9990
1	1.376	1.963	3.078	6.314	12.706	31.821	63.657	318.31	636.62
2	1.061	1.386	1.886	2.920	4.303	6.965	9.925	22.327	31.599
3	0.978	1.250	1.638	2.353	3.182	4.541	5.841	10.215	12.924
4	0.941	1.190	1.533	2.132	2.776	3.747	4.604	7.173	8.610
5	0.920	1.156	1.476	2.015	2.571	3.365	4.032	5.893	6.869
6	0.906	1.134	1.440	1.943	2.447	3.143	3.707	5.208	5.959
7	0.896	1.119	1.415	1.895	2.365	2.998	3.499	4.785	5.408
8	0.889	1.108	1.397	1.860	2.306	2.896	3.355	4.501	5.041
9	0.883	1.100	1.383	1.833	2.262	2.821	3.250	4.297	4.781
10	0.879	1.093	1.372	1.812	2.228	2.764	3.169	4.144	4.587
11	0.876	1.088	1.363	1.796	2.201	2.718	3.106	4.025	4.437
12	0.873	1.083	1.356	1.782	2.179	2.681	3.055	3.930	4.318
13	0.870	1.079	1.350	1.771	2.160	2.650	3.012	3.852	4.221
14	0.868	1.076	1.345	1.761	2.145	2.624	2.977	3.787	4.140
15	0.866	1.074	1.341	1.753	2.131	2.602	2.947	3.733	4.073
16	0.865	1.071	1.337	1.746	2.120	2.583	2.921	3.686	4.015
17	0.863	1.069	1.333	1.740	2.110	2.567	2.898	3.646	3.965
18	0.862	1.067	1.330	1.734	2.101	2.552	2.878	3.610	3.922
19	0.861	1.066	1.328	1.729	2.093	2.539	2.861	3.579	3.883
20	0.860	1.064	1.325	1.725	2.086	2.528	2.845	3.552	3.850
21	0.859	1.063	1.323	1.721	2.080	2.518	2.831	3.527	3.819
22	0.858	1.061	1.321	1.717	2.074	2.508	2.819	3.505	3.792
23	0.858	1.060	1.319	1.714	2.069	2.500	2.807	3.485	3.768
24	0.857	1.059	1.318	1.711	2.064	2.492	2.797	3.467	3.745
25	0.856	1.058	1.316	1.708	2.060	2.485	2.787	3.450	3.725
26	0.856	1.058	1.315	1.706	2.056	2.479	2.779	3.435	3.707
27	0.855	1.057	1.314	1.703	2.052	2.473	2.771	3.421	3.690
28	0.855	1.056	1.313	1.701	2.048	2.467	2.763	3.408	3.674
29	0.854	1.055	1.311	1.699	2.045	2.462	2.756	3.396	3.659
30	0.854	1.055	1.310	1.697	2.042	2.457	2.750	3.385	3.646

		Selected right-tail areas with confidence levels underneath							
conf lev	*0.2000*	*0.1500*	*0.1000*	*0.0500*	*0.0250*	*0.0100*	*0.0050*	*0.0010*	*0.0005*
df	*0.6000*	*0.7000*	*0.8000*	*0.9000*	*0.9500*	*0.9800*	*0.9900*	*0.9980*	*0.9990*
40	0.851	1.050	1.303	1.684	2.021	2.423	2.704	3.307	3.551
50	0.849	1.047	1.299	1.676	2.009	2.403	2.678	3.261	3.496
75	0.846	1.044	1.293	1.665	1.992	2.377	2.643	3.202	3.425
100	0.845	1.042	1.290	1.660	1.984	2.364	2.626	3.174	3.390
200	0.843	1.039	1.286	1.653	1.972	2.345	2.601	3.131	3.340
1000	0.842	1.037	1.282	1.646	1.962	2.330	2.581	3.098	3.300

Table 5 provides the chi-square statistic for the corresponding value of alpha and the number of degrees of freedom.

Table 5 Chi-Square Probability Distribution

			Selected right-tail areas					
df	*0.3000*	*0.2000*	*0.1500*	*0.1000*	*0.0500*	*0.0250*	*0.0100*	*0.0050*	*0.0010*
1	1.074	1.642	2.072	2.706	3.841	5.024	6.635	7.879	10.828
2	2.408	3.219	3.794	4.605	5.991	7.378	9.210	10.597	13.816
3	3.665	4.642	5.317	6.251	7.815	9.348	11.345	12.838	16.266
4	4.878	5.989	6.745	7.779	9.488	11.143	13.277	14.860	18.467
5	6.064	7.289	8.115	9.236	11.070	12.833	15.086	16.750	20.515
6	7.231	8.558	9.446	10.645	12.592	14.449	16.812	18.548	22.458
7	8.383	9.803	10.748	12.017	14.067	16.013	18.475	20.278	24.322
8	9.524	11.030	12.027	13.362	15.507	17.535	20.090	21.955	26.124
9	10.656	12.242	13.288	14.684	16.919	19.023	21.666	23.589	27.877
10	11.781	13.442	14.534	15.987	18.307	20.483	23.209	25.188	29.588
11	12.899	14.631	15.767	17.275	19.675	21.920	24.725	26.757	31.264
12	14.011	15.812	16.989	18.549	21.026	23.337	26.217	28.300	32.909
13	15.119	16.985	18.202	19.812	22.362	24.736	27.688	29.819	34.528
14	16.222	18.151	19.406	21.064	23.685	26.119	29.141	31.319	36.123
15	17.322	19.311	20.603	22.307	24.996	27.488	30.578	32.801	37.697
16	18.418	20.465	21.793	23.542	26.296	28.845	32.000	34.267	39.252
17	19.511	21.615	22.977	24.769	27.587	30.191	33.409	35.718	40.790

continues

Table 5 Chi-Square Probability Distribution (continued)

Selected right-tail areas

18	20.601	22.760	24.155	25.989	28.869	31.526	34.805	37.156	42.312
19	21.689	23.900	25.329	27.204	30.144	32.852	36.191	38.582	43.820
20	22.775	25.038	26.498	28.412	31.410	34.170	37.566	39.997	45.315
21	23.858	26.171	27.662	29.615	32.671	35.479	38.932	41.401	46.797
22	24.939	27.301	28.822	30.813	33.924	36.781	40.289	42.796	48.268
23	26.018	28.429	29.979	32.007	35.172	38.076	41.638	44.181	49.728
24	27.096	29.553	31.132	33.196	36.415	39.364	42.980	45.559	51.179
25	28.172	30.675	32.282	34.382	37.652	40.646	44.314	46.928	52.620
26	29.246	31.795	33.429	35.563	38.885	41.923	45.642	48.290	54.052
27	30.319	32.912	34.574	36.741	40.113	43.195	46.963	49.645	55.476
28	31.391	34.027	35.715	37.916	41.337	44.461	48.278	50.993	56.892
29	32.461	35.139	36.854	39.087	42.557	45.722	49.588	52.336	58.301
30	33.530	36.250	37.990	40.256	43.773	46.979	50.892	53.672	59.703

Table 6 provides the F-statistic for the corresponding degrees of freedom v_1 and v_2 using a value of alpha equal to 0.05.

Table 6 F-Distribution

$\alpha = 0.05$

\ v_1 v_2	1	2	3	4	5	6	7	8	9	10
1	161.448	199.500	215.707	224.583	230.162	233.986	236.768	238.882	240.543	241.882
2	18.513	19.000	19.164	19.247	19.296	19.330	19.353	19.371	19.385	19.396
3	10.128	9.552	9.277	9.117	9.013	8.941	8.887	8.845	8.812	8.786
4	7.709	6.944	6.591	6.388	6.256	6.163	6.094	6.041	5.999	5.964
5	6.608	5.786	5.409	5.192	5.050	4.950	4.876	4.818	4.772	4.735
6	5.987	5.143	4.757	4.534	4.387	4.284	4.207	4.147	4.099	4.060
7	5.591	4.737	4.347	4.120	3.972	3.866	3.787	3.726	3.677	3.637
8	5.318	4.459	4.066	3.838	3.687	3.581	3.500	3.438	3.388	3.347
9	5.117	4.256	3.863	3.633	3.482	3.374	3.293	3.230	3.179	3.137
10	4.965	4.103	3.708	3.478	3.326	3.217	3.135	3.072	3.020	2.978
11	4.844	3.982	3.587	3.357	3.204	3.095	3.012	2.948	2.896	2.854

$\alpha = 0.05$									
$\backslash v_1$ 1	2	3	4	5	6	7	8	9	10
v_2									
12 4.747	3.885	3.490	3.259	3.106	2.996	2.913	2.849	2.796	2.753
13 4.667	3.806	3.411	3.179	3.025	2.915	2.832	2.767	2.714	2.671
14 4.600	3.739	3.344	3.112	2.958	2.848	2.764	2.699	2.646	2.602
15 4.543	3.682	3.287	3.056	2.901	2.790	2.707	2.641	2.588	2.544
16 4.494	3.634	3.239	3.007	2.852	2.741	2.657	2.591	2.538	2.494
17 4.451	3.592	3.197	2.965	2.810	2.699	2.614	2.548	2.494	2.450
18 4.414	3.555	3.160	2.928	2.773	2.661	2.577	2.510	2.456	2.412
19 4.381	3.522	3.127	2.895	2.740	2.628	2.544	2.477	2.423	2.378
20 4.351	3.493	3.098	2.866	2.711	2.599	2.514	2.447	2.393	2.348
21 4.325	3.467	3.072	2.840	2.685	2.573	2.488	2.420	2.366	2.321
22 4.301	3.443	3.049	2.817	2.661	2.549	2.464	2.397	2.342	2.297
23 4.279	3.422	3.028	2.796	2.640	2.528	2.442	2.375	2.320	2.275
24 4.260	3.403	3.009	2.776	2.621	2.508	2.423	2.355	2.300	2.255
25 4.242	3.385	2.991	2.759	2.603	2.490	2.405	2.337	2.282	2.236
26 4.225	3.369	2.975	2.743	2.587	2.474	2.388	2.321	2.265	2.220
27 4.210	3.354	2.960	2.728	2.572	2.459	2.373	2.305	2.250	2.204
28 4.196	3.340	2.947	2.714	2.558	2.445	2.359	2.291	2.236	2.190
29 4.183	3.328	2.934	2.701	2.545	2.432	2.346	2.278	2.223	2.177
30 4.171	3.316	2.922	2.690	2.534	2.421	2.334	2.266	2.211	2.165
31 4.160	3.305	2.911	2.679	2.523	2.409	2.323	2.255	2.199	2.153
32 4.149	3.295	2.901	2.668	2.512	2.399	2.313	2.244	2.189	2.142
33 4.139	3.285	2.892	2.659	2.503	2.389	2.303	2.235	2.179	2.133
34 4.130	3.276	2.883	2.650	2.494	2.380	2.294	2.225	2.170	2.123
35 4.121	3.267	2.874	2.641	2.485	2.372	2.285	2.217	2.161	2.114
36 4.113	3.259	2.866	2.634	2.477	2.364	2.277	2.209	2.153	2.106
37 4.105	3.252	2.859	2.626	2.470	2.356	2.270	2.201	2.145	2.098
38 4.098	3.245	2.852	2.619	2.463	2.349	2.262	2.194	2.138	2.091
39 4.091	3.238	2.845	2.612	2.456	2.342	2.255	2.187	2.131	2.084
40 4.085	3.232	2.839	2.606	2.449	2.336	2.249	2.180	2.124	2.077
41 4.079	3.226	2.833	2.600	2.443	2.330	2.243	2.174	2.118	2.071
42 4.073	3.220	2.827	2.594	2.438	2.324	2.237	2.168	2.112	2.065

continues

Table 6   F-Distribution   (continued)

α = 0.05

\ v_1 1	2	3	4	5	6	7	8	9	10
v_2									
43 4.067	3.214	2.822	2.589	2.432	2.318	2.232	2.163	2.106	2.059
44 4.062	3.209	2.816	2.584	2.427	2.313	2.226	2.157	2.101	2.054
45 4.057	3.204	2.812	2.579	2.422	2.308	2.221	2.152	2.096	2.049
46 4.052	3.200	2.807	2.574	2.417	2.304	2.216	2.147	2.091	2.044
47 4.047	3.195	2.802	2.570	2.413	2.299	2.212	2.143	2.086	2.039
48 4.043	3.191	2.798	2.565	2.409	2.295	2.207	2.138	2.082	2.035
49 4.038	3.187	2.794	2.561	2.404	2.290	2.203	2.134	2.077	2.030
50 4.034	3.183	2.790	2.557	2.400	2.286	2.199	2.130	2.073	2.026

α = 0.05

\ v_1 11	12	13	14	15	16	17	18	19	20
v_2									
1 242.983	243.906	244.690	245.364	245.950	246.464	246.918	247.323	247.686	248.013
2 19.405	19.413	19.419	19.424	19.429	19.433	19.437	19.440	19.443	19.446
3 8.763	8.745	8.729	8.715	8.703	8.692	8.683	8.675	8.667	8.660
4 5.936	5.912	5.891	5.873	5.858	5.844	5.832	5.821	5.811	5.803
5 4.704	4.678	4.655	4.636	4.619	4.604	4.590	4.579	4.568	4.558
6 4.027	4.000	3.976	3.956	3.938	3.922	3.908	3.896	3.884	3.874
7 3.603	3.575	3.550	3.529	3.511	3.494	3.480	3.467	3.455	3.445
8 3.313	3.284	3.259	3.237	3.218	3.202	3.187	3.173	3.161	3.150
9 3.102	3.073	3.048	3.025	3.006	2.989	2.974	2.960	2.948	2.936
10 2.943	2.913	2.887	2.865	2.845	2.828	2.812	2.798	2.785	2.774
11 2.818	2.788	2.761	2.739	2.719	2.701	2.685	2.671	2.658	2.646
12 2.717	2.687	2.660	2.637	2.617	2.599	2.583	2.568	2.555	2.544
13 2.635	2.604	2.577	2.554	2.533	2.515	2.499	2.484	2.471	2.459
14 2.565	2.534	2.507	2.484	2.463	2.445	2.428	2.413	2.400	2.388
15 2.507	2.475	2.448	2.424	2.403	2.385	2.368	2.353	2.340	2.328
16 2.456	2.425	2.397	2.373	2.352	2.333	2.317	2.302	2.288	2.276
17 2.413	2.381	2.353	2.329	2.308	2.289	2.272	2.257	2.243	2.230
18 2.374	2.342	2.314	2.290	2.269	2.250	2.233	2.217	2.203	2.191
19 2.340	2.308	2.280	2.256	2.234	2.215	2.198	2.182	2.168	2.155

$\alpha = 0.05$

$\backslash v_1$ 11	12	13	14	15	16	17	18	19	20
v_2									
20 2.310	2.278	2.250	2.225	2.203	2.184	2.167	2.151	2.137	2.124
21 2.283	2.250	2.222	2.197	2.176	2.156	2.139	2.123	2.109	2.096
22 2.259	2.226	2.198	2.173	2.151	2.131	2.114	2.098	2.084	2.071
23 2.236	2.204	2.175	2.150	2.128	2.109	2.091	2.075	2.061	2.048
24 2.216	2.183	2.155	2.130	2.108	2.088	2.070	2.054	2.040	2.027
25 2.198	2.165	2.136	2.111	2.089	2.069	2.051	2.035	2.021	2.007
26 2.181	2.148	2.119	2.094	2.072	2.052	2.034	2.018	2.003	1.990
27 2.166	2.132	2.103	2.078	2.056	2.036	2.018	2.002	1.987	1.974
28 2.151	2.118	2.089	2.064	2.041	2.021	2.003	1.987	1.972	1.959
29 2.138	2.104	2.075	2.050	2.027	2.007	1.989	1.973	1.958	1.945
30 2.126	2.092	2.063	2.037	2.015	1.995	1.976	1.960	1.945	1.932
31 2.114	2.080	2.051	2.026	2.003	1.983	1.965	1.948	1.933	1.920
32 2.103	2.070	2.040	2.015	1.992	1.972	1.953	1.937	1.922	1.908
33 2.093	2.060	2.030	2.004	1.982	1.961	1.943	1.926	1.911	1.898
34 2.084	2.050	2.021	1.995	1.972	1.952	1.933	1.917	1.902	1.888
35 2.075	2.041	2.012	1.986	1.963	1.942	1.924	1.907	1.892	1.878
36 2.067	2.033	2.003	1.977	1.954	1.934	1.915	1.899	1.883	1.870
37 2.059	2.025	1.995	1.969	1.946	1.926	1.907	1.890	1.875	1.861
38 2.051	2.017	1.988	1.962	1.939	1.918	1.899	1.883	1.867	1.853
39 2.044	2.010	1.981	1.954	1.931	1.911	1.892	1.875	1.860	1.846
40 2.038	2.003	1.974	1.948	1.924	1.904	1.885	1.868	1.853	1.839
41 2.031	1.997	1.967	1.941	1.918	1.897	1.879	1.862	1.846	1.832
42 2.025	1.991	1.961	1.935	1.912	1.891	1.872	1.855	1.840	1.826
43 2.020	1.985	1.955	1.929	1.906	1.885	1.866	1.849	1.834	1.820
44 2.014	1.980	1.950	1.924	1.900	1.879	1.861	1.844	1.828	1.814
45 2.009	1.974	1.945	1.918	1.895	1.874	1.855	1.838	1.823	1.808
46 2.004	1.969	1.940	1.913	1.890	1.869	1.850	1.833	1.817	1.803
47 1.999	1.965	1.935	1.908	1.885	1.864	1.845	1.828	1.812	1.798
48 1.995	1.960	1.930	1.904	1.880	1.859	1.840	1.823	1.807	1.793
49 1.990	1.956	1.926	1.899	1.876	1.855	1.836	1.819	1.803	1.789
50 1.986	1.952	1.921	1.895	1.871	1.850	1.831	1.814	1.798	1.784

Glossary

Addition Rule of Probabilities   Determines the probability of the union of two or more events.

Alternative Hypothesis   Denoted by H_1, represents the opposite of the null hypothesis and holds true if the null hypothesis is found to be false.

Analysis of Variance (ANOVA)   A procedure to test the difference between more than two population means.

Bar Charts   A data display where the value of the observation is proportional to the height of the bar on the graph.

Bayes' Theorem   A theorem used to calculate P[B/A] from information about P[A/B]. The term P[A/B] refers to the probability of Event A, given that Event B has occurred.

Biased Sample   A sample that does not represent the intended population and can lead to distorted findings.

Binomial Experiment   An experiment that has only two possible outcomes for each trial. The probability of success and failure is constant. Each trial of the experiment is independent of any other trial.

Binomial Probability Distribution   Used to calculate the probability of a specific number of successes for a certain number of trials.

Central Limit Theorem   A theorem that states as the sample size, n, gets larger, the sample means tend to follow a normal probability distribution.

Classes   The intervals in a frequency distribution.

Classical Probability   Refers to situations when we know the number of possible outcomes of the event of interest.

Cluster Sample   A simple random sample of groups, or clusters, of the population. Each member of the chosen clusters would be part of the final sample.

Coefficient of Determination, r^2   Represents the percentage of the variation in y that is explained by the regression line.

Combinations   The number of different ways in which objects can be arranged without regard to order.

Completely Randomized One-Way ANOVA   An analysis of variance procedure that involves the independent random selection of observations for each level of one factor.

Conditional Probability   The probability of Event A knowing that Event B has already occurred.

Confidence Interval   A range of values used to estimate a population parameter and is associated with a specific confidence level.

Confidence Level   The probability that the interval estimate will include the population parameter.

Contingency Tables   Shows the actual or relative frequency of two types of data at the same time in a table.

Continuous Random Variable   A variable that can assume any numerical value within an interval as a result of measuring the outcome of an experiment.

Correlation Coefficient   Indicates the strength and direction of the linear relationship between the independent and dependent variables.

Cumulative Frequency Distribution   Indicates the percentage of observations that are less than or equal to the current class.

Data   The value assigned to an observation or a measurement and is the building block to statistical analysis.

Degrees of Freedom   The number of values that are free to be varied given information, such as the sample mean, is known.

Dependent Samples   The observation from one sample is related to an observation from another sample.

Dependent Variable   The variable denoted by y in the regression equation that is suspected to be influenced by the independent variable.

Descriptive Statistics   Used to summarize or display data so that we can quickly obtain an overview.

Direct Observation   Gathering data while the subjects of interest are in their natural environment.

Discrete Probability Distribution   A listing of all the possible outcomes of an experiment for a discrete random variable along with the relative frequency or probability.

Discrete Random Variable   A variable that is limited to assuming only specific integer values as a result of counting the outcome of an experiment.

Empirical Probability   Type of probability that observes the number of occurrences of an event through an experiment and calculates the probability from a relative frequency distribution.

Empirical Rule   If a distribution follows a bell-shaped, symmetrical curve centered around the mean, we would expect approximately 68, 95, and 99.7 percent of the values to fall within 1, 2, and 3 standard deviations around the mean respectively.

Expected Frequencies   The number of observations that would be expected for each category of a frequency distribution, assuming the null hypothesis is true with chi-squared analysis.

Experiment   The process of measuring or observing an activity for the purpose of collecting data.

Event   One or more outcomes that are of interest for the experiment and which is/are a subset of the sample space.

Factor   Describes the cause of the variation in the data for analysis of variance.

Frequency Distribution   A table that shows the number of data observations that fall into specific intervals.

Focus Groups   An observational technique where the subjects are aware that data is being collected. These are used by businesses to gather information in a group setting that is controlled by a moderator.

Fundamental Counting Principle   States if one event can occur in m ways and a second event can occur in n ways, the total number of ways both events can occur together is $m*n$ ways.

Goodness-of-Fit Test   Uses a sample to test whether a frequency distribution fits the predicted distribution.

Histogram   A bar graph showing the number of observations in each class as the height of each bar.

Hypothesis   An assumption about a population parameter.

Independent Events   The occurrence of Event B has no effect on the probability of Event A.

Independent Samples   The observation from one sample is not related to any observations from another sample.

Independent Variable   The variable that is denoted by x in the regression equation that is suspected to influence the dependent variable.

Inferential Statistics   Used to make claims or conclusions about a population based on a sample of data from that population.

Interquartile Range   Measures the spread of the center half of the data set and is used to identify outliers.

Intersection   Two or more events occurring at the same time.

Interval Estimate   Provides a range of values that best describe the population.

Interval Level of Measurement   Level of data that allows the use of addition and subtraction when comparing values, but the zero point is arbitrary.

Joint Probability   The probability of the intersection of two events.

Law of Large Numbers   States that when an experiment is conducted a large number of times, the empirical probabilities of the process will converge to the classical probabilities.

Least Squares Method   A mathematical procedure to identify the linear equation that best fits a set of ordered pairs by finding values for a, the y-intercept, and b, the slope. The goal of the least squares method is to minimize the total squared error between the values of y and $\hat{y}$.

Level   The number of categories within the factor of interest in the analysis of variance procedure.

Level of Significance (α)   Probability of making a Type I error.

Line Charts   A display where ordered pair data points are connected together with a line.

Margin of Error   Determines the width of a confidence interval and is calculated using $z_c \sigma_{\bar{x}}$.

Mean   Calculated by adding all the values in the data set and then dividing this result by the number of observations.

Mean Square Between (MSB)   A measure of variation between the sample means.

Mean Square Within (MSW)   A measure of variation within each sample.

Measure of Central Tendency   Describes the center point of our data set with a single value.

Measure of Relative Position   Describes the percentage of the data below a certain point.

Median   The value in the data set for which half the observations are higher and half the observations are lower.

Mode   The observation in the data set that occurs most frequently.

Multiplication Rule of Probabilities   Determines the probability of the intersection of two or more events.

Mutually Exclusive Events   When two events cannot occur at the same time during an experiment.

Nominal Level of Measurement   Lowest level of data where numbers are used to identify a group or category.

Null Hypothesis   Denoted by H_0, represents the status quo and involves stating the belief that the mean of the population is ≤, =, or ≥ a specific value.

Observed Frequencies   The number of actual observations noted for each category of a frequency distribution with chi-squared analysis.

Observed Level of Significance   The smallest level of significance at which the null hypothesis will be rejected, assuming the null hypothesis is true. It is also known as the p-value.

One-Tail Hypothesis Test   Is used when the alternative hypothesis is being stated as < or >.

One-Way ANOVA   Analysis of variance procedure where only one factor is being considered.

Ordinal Level of Measurement   Has all the properties of nominal data with the added feature that we can rank order the values from highest to lowest.

Outcome   A particular result of an experiment.

Outliers   Extreme values in a data set that should be discarded before analysis.

Parameter   Data that describes a characteristic about a population.

Percentiles   Measure the relative position of the data values by dividing the data set into 100 equal segments.

Permutations   The number of different ways in which objects can be arranged in order.

Pie Chart   Used to describe data from relative frequency distributions with a circle divided into portions whose area is equal to the relative frequency distribution.

Point Estimate   A single value that best describes the population of interest, the sample mean being the most common.

Poisson Probability Distribution   Used to calculate the probability that a certain number of events will occur over a specific period of time.

Pooled Estimate of the Standard Deviation   A weighted average of 2 sample variances.

Population   Represents all possible outcomes or measurements of interest.

Primary Data   Data that is collected by the person who eventually uses the data.

Probability   The likelihood that a particular event will occur.

Probability Distribution   A listing of all the possible outcomes of an experiment along with the relative frequency or probability of each outcome.

p-Value   The smallest level of significance at which the null hypothesis will be rejected, assuming the null hypothesis is true.

Qualitative Data   Uses descriptive terms to measure or classify something of interest.

Quantitative Data   Uses numerical values to describe something of interest.

Quartiles   Measures the relative position of the data values by dividing the data set into 4 equal segments.

Random Variable   A variable that takes on a numerical value as a result of an experiment.

Range   Obtained by subtracting the smallest measurement from the largest measurement from a sample.

Ratio Level of Measurement   Level of data that allows the use of all four mathematical operations to compare values and has a true zero point.

Relative Frequency Distribution   Displays the percentage of observations of each class relative to the total number of observations.

Sample   A subset of a population.

Sample Space   All the possible outcomes of an experiment.

Sampling Distribution for the Difference in Means   Describes the probability of observing various intervals for the difference between 2 sample means.

Sampling Distribution of the Mean   The pattern of the sample means that will occur as samples are drawn from the population at large.

Sampling Error   Occurs when the sample measurement is different from the population measurement.

Standard Error of the Difference between Two Means   Describes the variation in the difference between 2 sample means.

Standard Error of the Estimate, s_e   Measures the amount of dispersion of the observed data around the regression line.

Scheffé Test   Used to determine which of the sample means are different after rejecting the null hypothesis using analysis of variance.

Secondary Data   Data that somebody else has collected and makes it available for others to use.

Simple Random Sample   A sample where every element in the population has a chance at being selected.

Simple Regression   A procedure that describes a straight line that best fits a series of ordered pairs (x,y).

Standard Deviation   A measure of variation calculated by taking the square root of the variance.

Standard Error of the Mean   The standard deviation of sample means.

Standard Error of the Proportion   The standard deviation of the sample proportions.

Statistic   Data that describes a characteristic about a sample.

Statistics   The science that deals with the collection, tabulation, and systematic classification of quantitative data, especially as a basis for inference and induction.

Stem and Leaf Display   Displays the frequency distribution by splitting the data values into leaves (the last digit in the value) and stems (the remaining digits in the value).

Stratified Sample   A sample that is obtained by dividing the population into mutually exclusive groups, or strata, and randomly sampling from each of these groups.

Subjective Probability   Probabilities that are estimated based on experience and intuition.

Sum of Squares Between (SSB)   The variation among the samples in analysis of variance.

Sum of Squares Within (SSW)   The variation within the samples in analysis of variance.

Surveys   Data collection that involves directly asking the subject a series of questions.

Systematic Sample   A sample where every k^{th} member of the population is chosen for the sample, with value of k being approximately $\dfrac{N}{n}$, where N equals the size of the population and n equals the size of the sample.

Test Statistic   A quantity from a sample to be used to decide whether or not to reject the null hypothesis.

Total Sum of Squares   The total variation in analysis of variance which is obtained by adding the sum of squares between (SSB) and the sum of squares within (SSW).

Two-Tail Hypothesis Test   Is used whenever the alternative hypothesis is expressed as $\neq$.

Type I Error   Occurs when the null hypothesis is rejected when, in reality, it is true.

Type II Error   Occurs when the null hypothesis is accepted when, in reality, it is not true.

Union   At least one of a number of possible events occur.

Variance   A measure of dispersion that describes the relative distance between the data points in the set and the mean of the data set.

Weighted Mean   Allows the assignment of more weight to certain values and less weight to others when calculating an average.

Index